John D. Clayton & Peter W. Chung

Applied Finite Element Methods

Lecture Notes on Principles and Procedures

ISBN-10: 1721867465
ISBN-13: 978-1721867462
Printed by CreateSpace.

Printed in the United States of America.

Third printing (with corrections), August 2019.

Preface

This publication emerged out of lecture notes used in a one-semester course on Applied Finite Element Methods at the A. James Clark School of Engineering at the University of Maryland, College Park, Maryland, USA. While referred to throughout as a "book" or "textbook", this publication consists of course notes, computer examples, and problem sets converted to manuscript format. As such, the presentation in much of the text is rather informal, and the figures, while considered adequate for the current purpose, have not been professionally rendered. The university course, listed jointly in graduate programs in Mechanical Engineering and Aerospace Engineering, has been taught each summer since 2012. Traditionally, the course material has been of high interest to Masters students and active professionals returning to the university on a part-time basis to obtain graduate degrees.

The primary purpose of this work is to serve as a textbook for a first university course on the finite element method. The targeted student is a first-year graduate student in engineering—typically majoring in mechanical, aerospace, or civil engineering, or in engineering mechanics. Senior undergraduate students may also find the material accessible, depending on their educational background. The secondary purpose of this book is to serve as a desktop reference and learning tool for practicing engineers whose work involves numerical modeling and simulation.

The technical content can be categorized as follows. The introductory chapter provides general background information on finite element principles, with a focus on the steps involved in any practical finite element analysis of a real-world problem. Basic terminology and notation are introduced. The next four chapters consider finite element representations of time-independent, i.e, quasi-static, problems in mechanics and thermodynamics. Chapter 2 is focused on one-dimensional finite element analysis in engineering mechanics: truss and bar elements. Chapter 3 considers two- and three-dimensional problems involving beam and frame elements. Chapter 4 addresses planar problems in elasticity and heat transfer, i.e., two-dimensional finite element representations of boundary value problems in continuum mechanics. Chapter 5 deals with axisymmetric analysis. The final chapter, Chapter 6, describes dynamic or time-dependent events in the context of finite elements. A brief bibliography concludes each chapter, while a list of symbols, glossary, and subject index are included at the end of the book.

At times the presentation focuses on intuitive derivations based on physical concepts, often eschewing, but not contradicting, formal mathematics. Training in variational calculus and advanced continuum mechanics is not required of the reader/student. In this regard, the textbook is thought to be accessible to a broad audience, including working professionals for which possibly many years may have passed since completion of any prior coursework. In general, the mathematical complexity of the presentation tends to increase with increasing chapter number.

Most chapters contain, within the main text, example problems accompanied by solutions. As deemed appropriate for a book that may be used for instruction and reference, detailed derivations of important results are included throughout. Additional problems are provided at the end of each of the main Chapters 2 through 6 that may be used for student assignments or supplementary practice. Some of these problems can be solved analytically in closed form or with the aid of basic mathematical software. Brief solutions to many of these are listed in a separate section at the end of the book, just preceding the subject index. Other problems, particularly those at the end of later chapters, request application of (commercial) finite element software. By working through the problems within and at the end of each main chapter, students and practicing engineers will advance their knowledge of fundamental concepts that serve as a basis for modern finite element computer codes.

Also included as a distinct section at the end of each main chapter is an example problem solved using the finite element software package ANSYS[1]. In the presentation of each such example, key concepts are demonstrated in a step-by-step manner, with the intent that the general methods involved may be readily transferrable to any other commercial finite element code. The ANSYS Graphical User Interface (GUI), rather than the parametric design language, is used to treat each example.

This textbook is self-contained and can serve as a complete set of lecture notes for a one-semester course on finite element methods. The reader is assumed to have background knowledge of engineering mathematics—specifically including integral calculus and matrix algebra–as well as some training in strength of materials. Most of the text is directed towards mechanics of solids, though heat conduction is considered as well. Analogies to applications in other disciplines in physics and materials science are abundant; it follows that the text material may also be of use to students in numerous other branches of applied sciences.

The second author, Prof. Chung, designed the university course. The first author, Dr. Clayton, drafted the manuscript and therefore bears ultimate responsibility for any flaws, typesetting or otherwise, that it may contain. Errors can be reported to `jdclayt1@umd.edu` and will be corrected in future printings.

Aberdeen, Maryland, USA — *John D. Clayton*
College Park, Maryland, USA — *Peter W. Chung*

June 2018

[1] Specifically, ANSYS Mechanical Release 17.2, SAS IP, Inc., 2016.

Contents

Chapter 1
Introduction

Abstract A general description of the finite element method is provided along with requisite terminology. Representative features of mechanics boundary value problems are outlined. Steps involved in a finite element computer simulation are described.

1.1 The Finite Element Method: Overview

Finite Element Analysis (FEA) is a numerical technique for solving differential equation(s) whereby the solution domain is divided or discretized into one or more finite elements. With each element is associated a set of points called nodes, where the exact value of the solution, or some suitably accurate approximation to it, is sought. Within each element, at spatial locations distinct from nodal positions, the solution is interpolated via a predefined set of shape functions. The Finite Element Method (FEM) loosely refers to the steps involved in any application of FEA, and vice-versa.

The utility of FEA in engineering practice arises from the difficulty of solving most differential equations analytically. Such difficulties, often resulting from nonlinearities, may arise from complexities of the geometric domain, physical balance laws, nonlinear material behaviors, and/or coupling(s) among field variables. The FEM essentially enables transformation of a set of differential equations to a set of algebraic equations that can be solved readily using a digital computer. Modern FEA software also contains a number of features useful for engineering analysis and design: graphical interfaces for pre- and post-processing, various numerical solvers for efficient treatment of particular kinds of linear algebra and time integration problems, and code modules enabling consideration of multi-physics, for example.

Some key characteristics of FEM are illustrated in Fig. 1.1. Ongoing advances in FEA technology involve research communities in numerous branches of mathematics, physics, engineering, and computer science. Specifically, FEA historically has a mathematical basis in variational calculus, a physical basis in continuum field

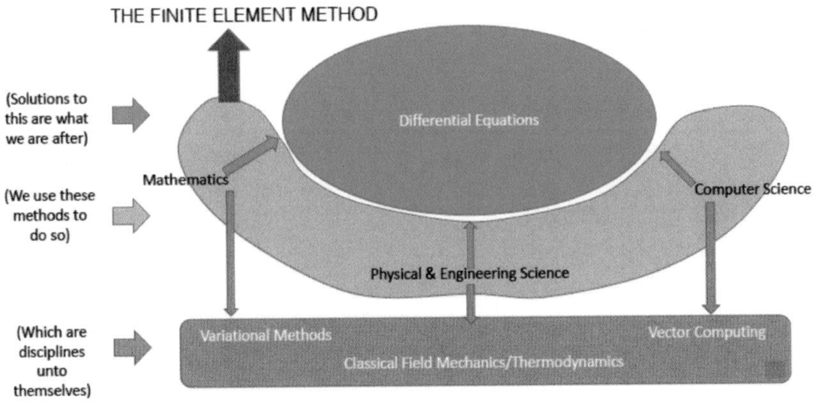

Fig. 1.1 The Finite Element Method (FEM): underlying scientific disciplines

theory, and a computer science basis in vector computing. The end goal in any case is solution of one or more differential equations representative of a true physical system.

1.2 Boundary Value Problems in the Physical Sciences

Before proceeding with a description of steps involved in FEA, a brief review of fundamental features in boundary value problems in field theories of continuous media is essential. Such features are universal to problems in various domains of physical science, including, but not limited to,

- Structural mechanics;
- Continuum mechanics of solids;
- Heat transfer;
- Fluid mechanics;
- Electromagnetics;
- Mass transport and diffusion.

Content in later chapters will focus on the first three physics domains—structural mechanics, solid mechanics (specifically elasticity), and heat transfer (specifically heat conduction)—but analogies to other branches of physical science arise frequently.

The aforementioned universal features consist of four main parts:

1. Compatibility relation: a mathematical relationship between primary variable(s) and secondary variable(s);
2. Conservation law: a governing equation from physics, typically a balance of force or energy;
3. Constitutive model: a mathematical representation of behavior of the particular substance or medium occupying the problem domain;
4. Boundary conditions: prescribed values of primary or secondary variables over some subset/boundary of the problem domain.

Quantities calculated directly in FEA are denoted as primary variables; quantities determined from primary variables are denoted as secondary variables. For example, in continuum mechanics of solids, displacement is a primary variable and strain and stress are secondary variables. Compatibility ensures that the strain tensor is the symmetric part of the spatial gradient of the displacement vector [1] in the small deformation case[1]. In heat transfer problems, the temperature is primary and the spatial temperature gradient and heat flux are secondary variables.

The conservation law(s) of a problem are the differential equations that are ultimately solved via FEA. For example, in mechanics of solids, the key governing equation is the balance of linear momentum. In heat conduction, the balance law is the First Law of Thermodynamics, i.e., the conservation of energy. A static (or quasi-static) problem is one in which time does not explicitly enter the conservation law, which in turn reduces to an equilibrium equation. In contrast, a dynamic problem includes time as an independent variable. An example of the latter in solid mechanics is acoustic wave propagation.

The constitutive model provides necessary information regarding material behavior and associated properties or parameters. In the case of elasticity, for example, the model is a relationship between the stress and strain, and the classical properties are the elastic moduli, e.g., the elastic constants in linear elasticity. In the case of heat conduction, the model is a relationship between the heat flux and temperature gradient, with the thermal conductivity the important material property. For dynamic problems, the mass density is also needed, as is the specific heat capacity for study of transient conduction.

Finally, boundary conditions distinguish one problem from another, and they enable a system of equations to have a non-trivial solution. Depending on the particular set of governing equations and constitutive model, this solution may or may not be unique.

In summary, one may remember the key parts defining a physical science problem as the "three C's plus BC's":

Compatibility, Conservation, Constitutive Model, Boundary Conditions

[1] In finite strain theory, compatibility requires that the deformation gradient tensor be the material gradient of the motion field [2, 3].

We will return to these fundamental characteristics often throughout the remainder of this textbook as each new class of problems is introduced in the context of FEA.

1.3 Finite Element Simulations

Any practical application of FEM towards solution of a physical problem using modern computer software can be broken down into a sequence of four key steps. The general sequence is

1. Define the problem.
2. Set up the problem.
3. Solve the problem.
4. Analyze the solution.

At the conclusion of step four, one may iterate on prior step(s) if the analysis is deemed unsatisfactory, for example if perceived errors in the numerical approximation are too large (e.g., due to a mesh that is too coarse), or if the result describes physically unrealistic behavior (e.g., due to a shortcoming in the constitutive model or associated material parameters).

Details regarding each step are provided in Fig. 1.2, and a complementary example is given in Fig. 1.3. The definition step consists of identification of the physical problem: geometry, behaviors of interest, and loading protocols. Overlapping the definition and setup steps is assignment of mathematical descriptors of the problem. An idealization of the geometry is often assumed, as in Fig. 1.3, where the rather complex shape of the body is approximated as circular in this example. The governing equation(s) and constitutive model(s) are assigned. A digital model is created, specifically a discretization of the domain into a mesh of finite elements. Boundary conditions are assigned to the mesh, and initial conditions are specified for time dependent problems. The FEM software then develops a set of algebraic equations corresponding to the governing equations and solves for primary field variables at the nodes. During the analysis step, the FEM software can be used to compute secondary variables from the primary solution. The user then may choose to judge the accuracy of the solution via comparison with data from another source, i.e., validation of the model results. This other source may be experimental data or an (approximate) analytical solution to the problem, if one exists.

As noted in Fig. 1.3, tasks performed by the computer software only address a small portion of the entire analysis (i.e., meshing, solving, and computing secondary variables); the problem definition, much of its setup, and thoughtful interpretation of the solution are responsibilities of the analyst or engineer. The former tasks performed by typical software are discussed next in a bit more detail.

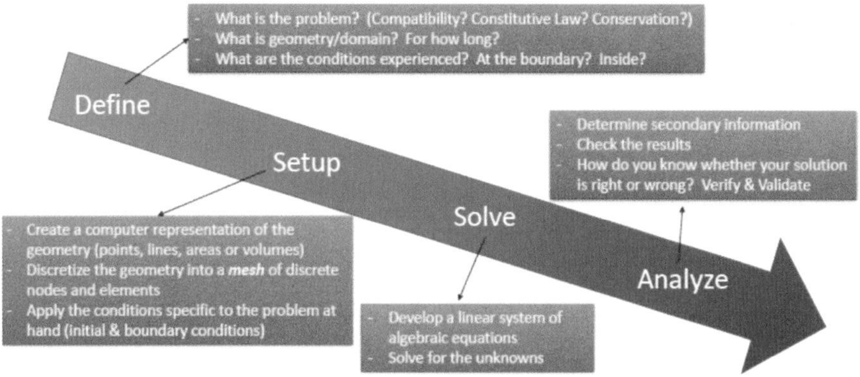
- What is the problem? (Compatibility? Constitutive Law? Conservation?)
- What is geometry/domain? For how long?
- What are the conditions experienced? At the boundary? Inside?
Define
Setup
Solve
Analyze
- Determine secondary information
- Check the results
- How do you know whether your solution is right or wrong? Verify & Validate
- Create a computer representation of the geometry (points, lines, areas or volumes)
- Discretize the geometry into a *mesh* of discrete nodes and elements
- Apply the conditions specific to the problem at hand (initial & boundary conditions)
- Develop a linear system of algebraic equations
- Solve for the unknowns

Fig. 1.2 Usual steps involved in finite element analysis

1.3.1 Pre-processing

The finite element pre-processor—invoked via a Graphical User Interface (GUI)—in any modern commercial FEM code, enables the analyst to create and mesh the problem domain, assign the governing equations, select material model(s) for constitutive behavior, and apply loading protocols.

The geometry of the problem is typically created by manipulation of points, lines, areas, and/or volumes. Various boolean operations such as adding, subtracting, and merging primitive shapes enable construction of complex geometries. Meshing involves selection of an element type or element types. Element types are logically classified as one-dimensional (1-D), two-dimensional (2-D), or three-dimensional (3-D). In general, the number of nodes in a polygonal or polyhedral element is equal to or greater than its number of vertices. Different element types of the same dimensionality have different numbers of nodes. As will be discussed in subsequent

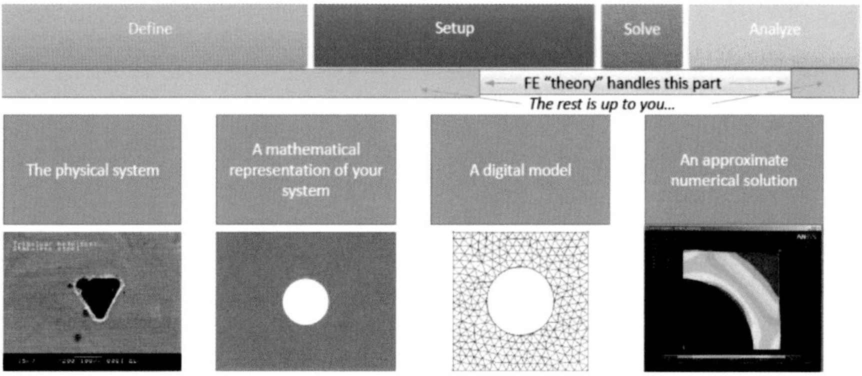
Define
Setup
Solve
Analyze
FE "theory" handles this part
The rest is up to you...
The physical system
A mathematical representation of your system
A digital model
An approximate numerical solution

Fig. 1.3 Example of steps involved in finite element analysis

chapters, the number of nodes is associated with the order of (polynomial) function used to interpolate the finite element solution within the element.

Most software packages enable a number of user options: free versus mapped meshing, mesh expansion, and mesh refinement. Refinement may be executed via adaptive h-refinement, which involves use of a finer mesh in areas of interest, or p-refinement, which entails use of increasingly higher-order elements. Other kinds of refinement are hp-refinement, which combines features of h- and p-refinement, and r-refinement, which involves strategically translating the mesh as the analysis proceeds incrementally.

Assignment of governing equations and material behavior is usually performed in tandem. For example, in a solid mechanics analysis, one might specify quasi-static equilibrium equations and elastic constitutive laws. Material properties are assigned that correspond to the chosen constitutive model(s). Again, in the example of linear elasticity, this would involve assignment of elastic constants relating stress to strain [1, 4]. Other more sophisticated models accounting for nonlinear behavior—nonlinear elasticity, viscoelasticity, plasticity, and so forth—are also available in most modern software packages.

Boundary conditions assigned to any part of the boundary of the domain may be of two usual types. Essential boundary conditions, also called Dirichlet boundary conditions, are constraints on or prescribed values of primary variables. Natural boundary conditions, also called Neumann boundary conditions, are constraints on or prescribed values of secondary variables. Mixed boundary conditions are also possible for some problems, wherein essential boundary conditions may be imposed for some degrees of freedom and natural boundary conditions for others at the same location on the body or its FE mesh.

1.3.2 Solving

The solution step yields values of primary field variables at nodes of the mesh. In the static problems we will encounter later, the equations to be solved can be written in the general form

$$[\boldsymbol{K}]\{\boldsymbol{u}\} = \{\boldsymbol{F}\}, \tag{1.1}$$

where $\boldsymbol{K}$ is a stiffness matrix, $\boldsymbol{u}$ is a vector of primary variables or degrees-of-freedom, and $\boldsymbol{F}$ is a load vector. The finite element solver essentially manipulates (1.1) into the following form and then obtains a solution numerically:

$$\{\boldsymbol{u}\} = [\boldsymbol{K}]^{-1}\{\boldsymbol{F}\}, \tag{1.2}$$

with $[\cdot]^{-1}$ denoting matrix inversion. Most stiffness matrices for the classes of problems addressed in this text are are both symmetric and sparse, and all we will encounter are positive-definite. As will be discussed in more detail in later chapters, various direct solvers and iterative solvers are available in commercial software packages. Direct solvers compute $[\boldsymbol{K}]^{-1}$ exactly, a procedure that becomes expensive for large systems. Iterative solvers seek approximate solutions within some prescribed tolerance, tending to compromise accuracy for wall-clock speed in obtaining the (approximate) solution.

1.3.3 Post-processing

Post-processing steps enabled by any complete and modern FEA software package enable visualization and analysis of the important outcomes of the model. Secondary variables can be computed and viewed as contours, for example, or can be exported as raw data to files for further manipulation in spreadsheets, etc. In a solid mechanics analysis, for example, typically one will seek to view stress fields in the body, where stress is a secondary variable computed from gradients of displacement, the primary variable. Modern software packages allow for animation as well, depicting deformed shapes of bodies as loads are applied incrementally. During post-processing, performance of an engineering design is often evaluated, e.g., does the system respond to applied loading as intended, within required tolerances or limits on stresses and deflections?

References

1. L.E. Malvern, *Introduction to the Mechanics of a Continuous Medium* (Prentice-Hall, Englewood Cliffs, NJ, 1969)
2. A.C. Eringen, *Nonlinear Theory of Continuous Media* (McGraw-Hill, NewYork, 1962)
3. J.D. Clayton, *Differential Geometry and Kinematics of Continua* (World Scientific, Singapore, 2014)
4. J.D. Clayton, *Nonlinear Mechanics of Crystals* (Springer, Dordrecht, 2011)

Chapter 2
Trusses and Bars

Abstract The finite element method is developed for elements that support only axial loads: trusses and bars. A brief review of essential matrix operations from linear algebra is given. The Direct Method, based on intuitive physical notions, is described for deriving the matrix equations for a system of truss or bar elements. The Finite Element Method is then presented for truss and bar elements, including strong and weak form equations, shape functions, element equilibrium equations, and post-processing.

Chapter 2 develops the finite element method for 1-D elements, specifically truss and bar elements. Prior to presentation of FEA for these element types, the same matrix equations used in FEA are first derived in an intuitive fashion via the so-called Direct Method (DM). Presentation of the Direct Method provides a physical basis for derivations that follow in the more rigorous mathematical context of FEM, and it also enables verification that the equations derived via FEM are physically realistic via their coincidence (in certain circumstances) with those derived via DM. Before proceeding with either method, a brief review of some identities from linear algebra is in order.

2.1 Linear Algebra

A number of simple identities that will be used frequently in future derivations and examples involving matrix operations are listed below. Explicit formulae are given for three 2×2 matrices and one column vector:

$$[\boldsymbol{A}] = \begin{bmatrix} a & b \\ c & d \end{bmatrix}, \quad [\boldsymbol{B}] = \begin{bmatrix} m & n \\ p & q \end{bmatrix}, \quad [\boldsymbol{C}] = \begin{bmatrix} r & s \\ t & w \end{bmatrix}; \qquad \{\boldsymbol{v}\} = \begin{Bmatrix} x \\ y \end{Bmatrix}. \tag{2.1}$$

The typical convention used here is enclosure of matrices in square brackets ($[\cdot]$) and vectors in braces ($\{\cdot\}$), though such enclosures may be omitted at times. Bold font is used for representation of matrices and vectors, while individual entries of matrices, vectors, and general tensors (i.e., all scalars) are written in italic font. Greek fonts will also be used often for scalars.

The following formulae apply, where all entries are assumed to be real numbers:

- Addition

$$[\boldsymbol{A}]+[\boldsymbol{B}]=\begin{bmatrix} a+m & b+n \\ c+p & d+q \end{bmatrix}; \tag{2.2}$$

- Multiplication by a scalar

$$\alpha[\boldsymbol{A}]=\begin{bmatrix} \alpha a & \alpha b \\ \alpha c & \alpha d \end{bmatrix}; \tag{2.3}$$

- Linearity

$$\alpha[\boldsymbol{A}][\boldsymbol{B}]=[\boldsymbol{A}]\alpha[\boldsymbol{B}]; \tag{2.4}$$

- Linear combination

$$\alpha[\boldsymbol{A}]+\beta[\boldsymbol{B}]=\begin{bmatrix} \alpha a+\beta m & \alpha b+\beta n \\ \alpha c+\beta p & \alpha d+\beta q \end{bmatrix}; \tag{2.5}$$

- Matrix-vector multiplication

$$[\boldsymbol{A}]\{\boldsymbol{v}\}=\begin{bmatrix} a & b \\ c & d \end{bmatrix}\begin{Bmatrix} x \\ y \end{Bmatrix}=\begin{Bmatrix} ax+by \\ cx+dy \end{Bmatrix}; \tag{2.6}$$

- Matrix multiplication

$$[\boldsymbol{A}][\boldsymbol{B}]=\begin{bmatrix} a & b \\ c & d \end{bmatrix}\begin{bmatrix} m & n \\ p & q \end{bmatrix}=\begin{bmatrix} am+bp & an+bq \\ cm+dp & cn+dq \end{bmatrix}, \quad (\text{In general }, [\boldsymbol{A}][\boldsymbol{B}]\neq[\boldsymbol{B}][\boldsymbol{A}]); \tag{2.7}$$

- Matrix transposition

$$[\boldsymbol{A}]^{\mathrm{T}}=\begin{bmatrix} a & b \\ c & d \end{bmatrix}^{\mathrm{T}}=\begin{bmatrix} a & c \\ b & d \end{bmatrix}, \qquad ([\boldsymbol{A}][\boldsymbol{B}])^{\mathrm{T}}=[\boldsymbol{B}]^{\mathrm{T}}[\boldsymbol{A}]^{\mathrm{T}}; \tag{2.8}$$

- Matrix inversion

$$[\boldsymbol{C}]^{-1}=\frac{1}{rw-st}\begin{bmatrix} w & -s \\ -t & r \end{bmatrix}=\frac{1}{\det\boldsymbol{C}}\begin{bmatrix} w & -s \\ -t & r \end{bmatrix}, \quad [\boldsymbol{C}]^{-1}[\boldsymbol{C}]=[\boldsymbol{I}]=\begin{bmatrix} 1 & 0 \\ 0 & 1 \end{bmatrix}; \tag{2.9}$$

- Matrix determinant

$$\det\boldsymbol{C}=|\boldsymbol{C}|=rw-st, \qquad \det(\boldsymbol{C}^{-1})=\frac{1}{\det\boldsymbol{C}}, \qquad \det(\boldsymbol{C}^{\mathrm{T}})=\det\boldsymbol{C}. \tag{2.10}$$

The identity matrix is denoted by $[\boldsymbol{I}]$. A matrix whose determinant is zero is said to be singular and is not invertible. In contrast, an invertible matrix has a nonzero determinant and is said to be non-singular. A real matrix $\boldsymbol{C}$ is said to be symmetric if $\boldsymbol{C}^{\mathrm{T}} = \boldsymbol{C}$ and is said to be skew if $\boldsymbol{C}^{\mathrm{T}} = -\boldsymbol{C}$. A recommended reference for further study of these and other concepts of linear algebra is [1].

2.2 The Direct Method

Two numerical methods discussed and compared in Chapter 2 are the Direct Method (DM) and Finite Element Method (FEM). As indicated in Table 2.1, most features are identical across the two methods, the only major difference being the way in which the governing vector-matrix equations are derived.

Numerical methods apply advanced mathematics to recast a problem often difficult to solve analytically in a form amenable to more tractable solution via a digital computer. The tedious steps involved in the solution are thereby transferred from the analyst to the computer. Computers enable simple operations to be performed quickly and many times over. The simpler the effort for the computer, the faster its performance. The FEM breaks down an otherwise complicated problem into a series of repeatable arithmetic (i.e., addition, subtraction, multiplication, and division) operations. The most efficient FEA codes are those that handle book-keeping or data management tasks with the most alacrity since the underlying physics/engineering concepts and mathematics are identical for a given class of problems.

Table 2.1 Comparison of steps in DM and FEM

Step	Direct Method	Finite Element Method
Derivation	From physical arguments	From differential equations
Assembly	Same	Same
Boundary Conditions	Same	Same
Solution	Same	Same
Post-processing	Same	Same

2.2.1 Element Equations

We begin by first considering a single spring, which as will be shown later, can be represented by a truss or bar element. The spring in Fig. 2.1 is linear in force-displacement behavior, with stiffness constant k. Displacements and forces are assumed measurable at the two ends of the spring. Initially, the spring is undeformed, and subsequently it is loaded quasi-statically (i.e., no inertial effects and no damping). From basic Newtonian physics, static equilibrium requires that the net force on the spring be zero, i.e.,

$$Q_1 + Q_2 = 0, \tag{2.11}$$

Fig. 2.1 Linear spring

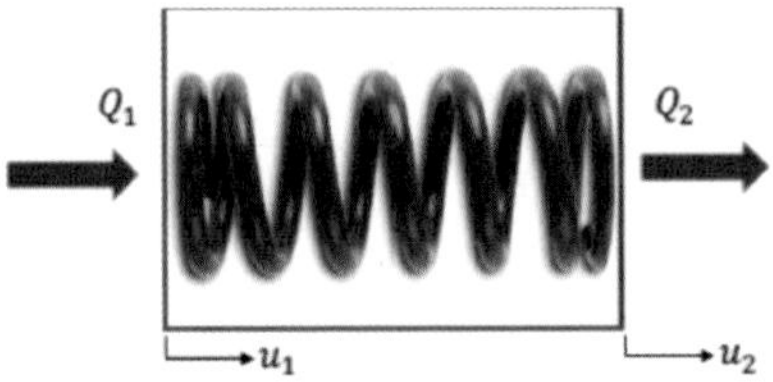

where, with u_1 and u_2 the endpoint displacements, the forces acting on the endpoints are

$$Q_1 = k(u_1 - u_2), \qquad Q_2 = k(u_2 - u_1). \tag{2.12}$$

The two equations in (2.12) can be recast in vector-matrix form as

$$k\begin{bmatrix} 1 & -1 \\ -1 & 1 \end{bmatrix}\begin{Bmatrix} u_1 \\ u_2 \end{Bmatrix} = \begin{Bmatrix} Q_1 \\ Q_2 \end{Bmatrix} \Leftrightarrow [\boldsymbol{K}]\{\boldsymbol{u}\} = \{\boldsymbol{Q}\}, \tag{2.13}$$

with $\boldsymbol{K}$, $\boldsymbol{u}$, and $\boldsymbol{Q}$ the stiffness matrix, displacement vector, and force vector, respectively.

As written in (2.13), displacements $\boldsymbol{u}$ cannot be obtained directly via solution of

$$\{\boldsymbol{u}\} = [\boldsymbol{K}]^{-1}\{\boldsymbol{Q}\} \tag{2.14}$$

because $\boldsymbol{K}$ is presently singular: its inverse is not defined since $\det\boldsymbol{K} = 0$. This problem is said to be statically indeterminant, which means that not enough information has been supplied to enable a solution. In this case, the indeterminancy arises because boundary conditions have not yet been applied. In general, the stiffness matrix for a given FEA problem may be singular if the problem is underconstrained, which often means rigid body modes of deformation are possible. For example, in this case, one could add a constant displacement of the same value to each of u_1 and u_2 without violating the static equilibrium requirement.

Now apply the following set of boundary conditions to the problem shown in Fig. 2.1:

$$u_1 = 0, \qquad Q_2 = P, \tag{2.15}$$

which impose a fixed left end of the spring and a tensile force of magnitude P acting on the right end. Substituting into algebraic equations in (2.12) gives

$$-ku_2 = Q_1, \quad ku_2 = P \quad \Rightarrow \quad u_2 = P/k, \quad Q_1 = -P. \tag{2.16}$$

Matrix equations (2.13) can be manipulated using (2.15) to obtain the same solution:

$$k\begin{bmatrix} 1 & 0 \\ 0 & 1 \end{bmatrix}\begin{Bmatrix} 0 \\ u_2 \end{Bmatrix} = \begin{Bmatrix} 0 \\ P \end{Bmatrix} \Rightarrow \begin{Bmatrix} u_1 \\ u_2 \end{Bmatrix} = \frac{1}{k}\begin{bmatrix} 1 & 0 \\ 0 & 1 \end{bmatrix}\begin{Bmatrix} 0 \\ P \end{Bmatrix} \Rightarrow u_2 = \frac{P}{k}, \tag{2.17}$$

$$\begin{Bmatrix} Q_1 \\ Q_2 \end{Bmatrix} = k \begin{bmatrix} 1 & -1 \\ -1 & 1 \end{bmatrix} \begin{Bmatrix} 0 \\ P/k \end{Bmatrix} = \begin{Bmatrix} -P \\ P \end{Bmatrix} \Rightarrow Q_1 = -P. \tag{2.18}$$

Note that in (2.17), the "effective" stiffness matrix $k\boldsymbol{I}$ is now invertible, and the problem is statically determinant. The substitution step in (2.18) of the primary solution variable u_2 into the original equation to obtain the secondary variable Q_1 is an example of post-processing.

As shown in Fig. 2.2, a spring can be used to represent a truss or bar with Young's modulus E, cross-sectional area A, and length L. When the product of E and A is constant over the entire length L, the effective spring-like stiffness of the bar or truss is

$$k = \frac{EA}{L} \qquad (EA = \text{constant}). \tag{2.19}$$

Here, a truss or bar element is defined as a mechanical object that supports only axial forces, i.e., a 1-D structural object. In later chapters, beam and frame elements that can support loads in one or more transverse directions will be introduced.

The problem of an axially loaded linear spring is simple enough that the governing equations can be obtained by inspection and solved using high school algebra. In this case, the algebraic approach is easier than use of vector-matrix representations or FEA. In contrast, when a more complicated structure is considered requiring representation by more than one element, the equations become unwieldy for basic algebra, and vector-matrix representations become essential.

2.2.2 Assembly

The assembly process is necessary for problems consisting of more than one element. When more than one element is present, a numbering system must be established to keep track of nodes and elements. Different textbooks and different FEM codes have different schemes. The convention used in this book is illustrated in Fig. 2.3.

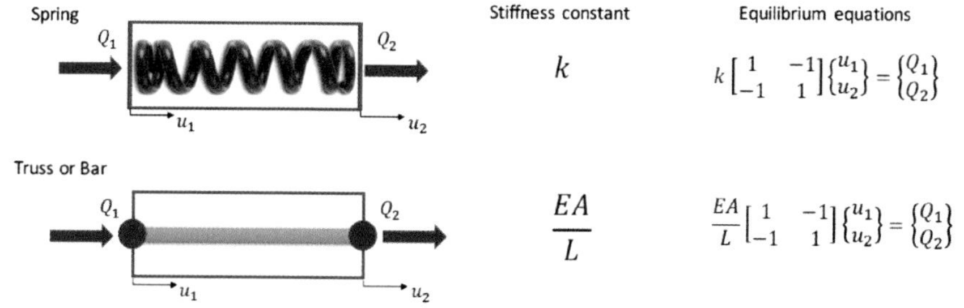

Fig. 2.2 1-D mechanical elements: spring and truss or bar

Quantities associated with local node numbers within an element e are denoted with a superscript in parentheses for that element. Element number e runs from 1 to the total number of elements in the mesh. The subscript on the quantity denotes the local node number for that particular element; such a subscript runs from 1 to the number of nodes contained in a single element. The same node may have a different local number depending on the element from which the node is observed. In other words, a node shared among multiple elements may have a different local number for each element.

Quantities associated with global node numbers have no superscript. Global numbers, which are denoted by subscripts, are referenced after all of the elements in the mesh are assembled. Global node numbers run from 1 to the total number of nodes in the whole mesh.

Consider again the two-element example in Fig. 2.3. The governing static equilibrium equations in algebraic form are expressed as follows:

$$ku_1^{(1)} - ku_2^{(1)} = Q_1^{(1)}, \qquad -ku_1^{(1)} + ku_2^{(1)} = Q_2^{(1)} \quad \text{(element 1)}; \tag{2.20}$$

$$ku_1^{(2)} - ku_2^{(2)} = Q_1^{(2)}, \qquad -ku_1^{(2)} + ku_2^{(2)} = Q_2^{(2)} \quad \text{(element 2)}. \tag{2.21}$$

The following notation and conditions relating local and global variables (displacement and force) are noted:

$$u_1^{(1)} = u_1, \qquad u_2^{(1)} = u_1^{(2)} = u_2, \qquad u_2^{(2)} = u_3; \tag{2.22}$$

$$Q_1^{(1)} = Q_1, \qquad Q_2 = Q_2^{(1)} + Q_1^{(2)} = 0, \qquad Q_2^{(2)} = Q_3. \tag{2.23}$$

Adding the second of (2.20) to the first of (2.21) and invoking relationships in (2.22) and (2.23), we arrive at a system of three algebraic equations and three global unknowns u_1, u_2, u_3:

$$k(u_1 - u_2) = Q_1, \quad k(-u_1 + 2u_2 - u_3) = 0, \quad k(-u_2 + u_3) = Q_3. \tag{2.24}$$

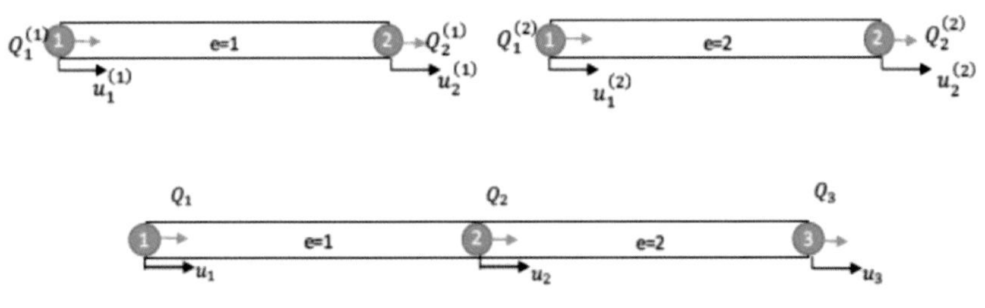

Fig. 2.3 Local (top) and global (bottom) numbering schemes for a two-element truss structure

If boundary conditions are then applied, the system can be solved. However the above algebraic approach is rather cumbersome, even for a relatively simple physical system consisting of only two serial linear elements.

In contrast, representation by vectors and matrices is much more efficient when multiple elements are involved. Equation (2.13) can be written for each element in the form

$$\begin{bmatrix} k & -k \\ -k & k \end{bmatrix} \begin{Bmatrix} u_1 \\ u_2 \end{Bmatrix} = \begin{Bmatrix} Q_1 \\ Q_2 \end{Bmatrix} \rightarrow \begin{bmatrix} k_{11}^{(e)} & k_{12}^{(e)} \\ k_{21}^{(e)} & k_{22}^{(e)} \end{bmatrix} \begin{Bmatrix} u_1^{(e)} \\ u_2^{(e)} \end{Bmatrix} = \begin{Bmatrix} Q_1^{(e)} \\ Q_2^{(e)} \end{Bmatrix}, \tag{2.25}$$

or symbolically as

$$[\boldsymbol{k}^{(e)}]\{\boldsymbol{u}^{(e)}\} = \{\boldsymbol{Q}^{(e)}\}. \tag{2.26}$$

The element stiffness matrix is written as $\boldsymbol{k}^{(e)}$. Element displacement and force vectors are denoted, respectively, by $\boldsymbol{u}^{(e)}$ and $\boldsymbol{Q}^{(e)}$. Combining such a result for two elements ($e = 1, 2$), we obtain a global stiffness matrix $\boldsymbol{K}$ and a global force vector $\boldsymbol{Q}$:

$$[\boldsymbol{K}] = \begin{bmatrix} K_{11} & K_{12} & K_{13} \\ K_{21} & K_{22} & K_{23} \\ K_{31} & K_{32} & K_{33} \end{bmatrix} = \begin{bmatrix} k_{11}^{(1)} & k_{12}^{(1)} & 0 \\ k_{21}^{(1)} & k_{22}^{(1)} + k_{11}^{(2)} & k_{12}^{(2)} \\ 0 & k_{21}^{(2)} & k_{22}^{(2)} \end{bmatrix}; \tag{2.27}$$

$$\{\boldsymbol{Q}\} = \begin{Bmatrix} Q_1 \\ Q_2 \\ Q_3 \end{Bmatrix} = \begin{Bmatrix} Q_1^{(1)} \\ Q_2^{(1)} + Q_1^{(2)} \\ Q_2^{(2)} \end{Bmatrix}. \tag{2.28}$$

Such assembly is written in short-hand or symbolic form as

$$[\boldsymbol{K}] = \bigwedge_{e=1,2} [\boldsymbol{k}^{(e)}]; \qquad \{\boldsymbol{Q}\} = \bigwedge_{e=1,2} \{\boldsymbol{Q}^{(e)}\}. \tag{2.29}$$

The operator $\bigwedge$ is used to denote assembly, a mathematical procedure which is obviously different than simple summation.

The following characteristic developments are noted. The dimensions of $\boldsymbol{K}$ and $\boldsymbol{Q}$ are $n \times n$ and n, respectively, where n is the number of nodes in the mesh, i.e., the maximum global node number if nodes are numbered consecutively starting from 1. Entries of global stiffness matrix $[\boldsymbol{K}]$ are of magnitude equal to those of the local stiffness for isolated nodes and a summation of contributions of local stiffness for shared nodes. The analogous discussion holds for entries of global force vector $\boldsymbol{Q}$, whose entries are the sum of all internal forces at each node. Another noteworthy characteristic is that, as will become more evident later in the context of larger physical systems, much of the global stiffness matrix for 1-D mechanical elements is of tridiagonal form, meaning apart from the diagonal and its neighboring left/right entries, most of the matrix components are of zero magnitude. Local and global stiffness matrices are all notably symmetric in this example, features which will hold for nearly all classes of problems addressed in this textbook.

Using (2.22) and (2.23) and making the substitutions

$$k_{11}^{(e)} = k_{22}^{(e)} = k, \quad k_{12}^{(e)} = k_{21}^{(e)} = -k, \tag{2.30}$$

the global equilibrium equation in vector-matrix form is

$$[\boldsymbol{K}]\{\boldsymbol{u}\} = \{\boldsymbol{Q}\} \to k \begin{bmatrix} 1 & -1 & 0 \\ -1 & 2 & -1 \\ 0 & -1 & 1 \end{bmatrix} \begin{Bmatrix} u_1 \\ u_2 \\ u_3 \end{Bmatrix} = \begin{Bmatrix} Q_1 \\ 0 \\ Q_3 \end{Bmatrix}. \tag{2.31}$$

Verification is straightforward that (2.31) is consistent with the algebraically derived system of equations in (2.24). However, in contrast to the former, the procedure for derivation of the equations in vector-matrix form is nearly universal for a given class of problems and associated element types, meaning that it can be automated and performed efficiently by a digital computer. In FEA, any modern software package executes such tasks rather seamlessly even for extremely large meshes containing many millions of nodes and elements. An example in the next few sections presents derivation, solution, and analysis of the governing equations for a somewhat larger system consisting of five elements.

2.2.3 An Example: Problem Setup

We consider a problem consisting of five linear springs as shown in Fig. 2.4. Two pairs of springs are loaded in parallel, with a single spring in between linking the two pairs. A force P is applied to the right end of the system, and the left end of the system is fixed to a wall, though later we will assign the boundary condition at this end in a general way as $u_1 = g$, with g prescribed. Each spring will later be assigned the same constant stiffness k, but the general analysis does not make this assertion until much later.

Each spring is represented by a single finite element with two nodes. The structure contains a total of four nodes, with a global displacement u_j, $j = 1, \ldots, 4$ associated with each node as illustrated in Fig. 2.4. Recall that the static equilibrium equation for a single spring element in vector-matrix form is

$$k \begin{bmatrix} 1 & -1 \\ -1 & 1 \end{bmatrix} \begin{Bmatrix} u_1 \\ u_2 \end{Bmatrix} = \begin{Bmatrix} Q_1 \\ Q_2 \end{Bmatrix} \Leftrightarrow [\boldsymbol{k}^{(e)}]\{\boldsymbol{u}^{(e)}\} = \{\boldsymbol{Q}^{(e)}\}. \tag{2.32}$$

When applied to element 1, this becomes

$$\begin{bmatrix} k_{11}^{(1)} & k_{12}^{(1)} \\ k_{21}^{(1)} & k_{22}^{(1)} \end{bmatrix} \begin{Bmatrix} u_1^{(1)} \\ u_2^{(1)} \end{Bmatrix} = \begin{Bmatrix} Q_1^{(1)} \\ Q_2^{(1)} \end{Bmatrix} \Leftrightarrow [\boldsymbol{k}^{(1)}]\{\boldsymbol{u}^{(1)}\} = \{\boldsymbol{Q}^{(1)}\}. \tag{2.33}$$

After constructing a free body diagram for each element as shown in Fig. 2.5, it is clear that a similar equation can be derived for each of the other four elements.

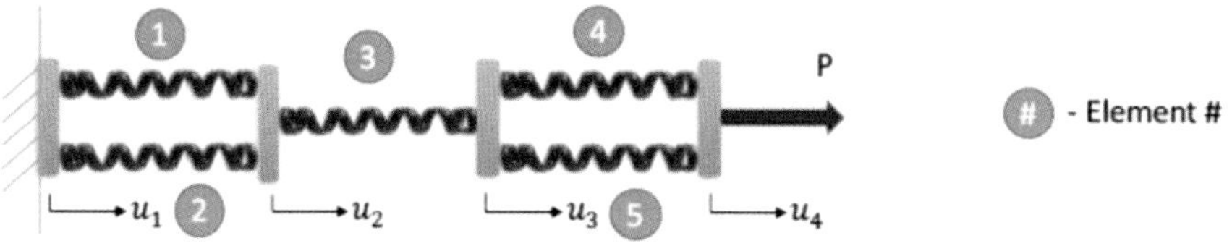

Fig. 2.4 Example problem setup: 5 springs

2.2.4 An Example: Assembly

Next, we consider the connectivity of the mesh, which provides information on how all of the local element matrices and vectors will be assembled into a global system of equations. Essential connectivity data is often organized in a connectivity table as shown for this example in Table 2.2. We see that node 1 is shared by elements 1 and 2; node 2 is shared by elements 1, 2, and 3; node 3 is shared by elements 3, 4, and 5; and node 4 is shared by elements 4 and 5. Mathematically, the connectivity of the mesh leads to the following constraints on nodal displacements:

$$\begin{aligned} u_1^{(1)} &= u_1^{(2)} = u_1, \qquad u_2^{(1)} = u_2^{(2)} = u_1^{(3)} = u_2, \\ u_2^{(3)} &= u_1^{(4)} = u_1^{(5)} = u_3, \qquad u_2^{(4)} = u_2^{(5)} = u_4. \end{aligned} \tag{2.34}$$

Use of (2.34) in conjunction with the equilibrium equations for each spring produces a system of four equations and four unknowns u_1, u_2, u_3, u_4:

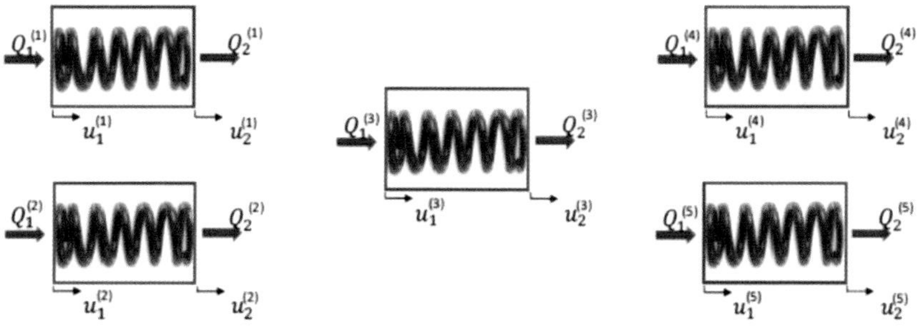

Fig. 2.5 Example problem: free body diagrams

$$
\begin{aligned}
&(k_{11}^{(1)}+k_{11}^{(2)})u_1+(k_{12}^{(1)}+k_{12}^{(2)})u_2=Q_1^{(1)}+Q_1^{(2)},\\
&(k_{21}^{(1)}+k_{21}^{(2)})u_1+(k_{22}^{(1)}+k_{22}^{(2)}+k_{11}^{(3)})u_2+k_{12}^{(3)}u_3=Q_2^{(1)}+Q_2^{(2)}+Q_1^{(3)},\\
&k_{21}^{(3)}u_3+(k_{22}^{(3)}+k_{11}^{(4)}+k_{11}^{(5)})u_4+(k_{12}^{(4)}+k_{12}^{(5)})u_5=Q_2^{(3)}+Q_1^{(4)}+Q_1^{(5)},\\
&(k_{21}^{(4)}+k_{21}^{(5)})u_4+(k_{22}^{(4)}+k_{22}^{(5)})u_5=Q_2^{(4)}+Q_2^{(5)}.
\end{aligned}
\tag{2.35}
$$

In vector-matrix form, this becomes

$$
\begin{bmatrix}
k_{11}^{(1)}+k_{11}^{(2)} & k_{12}^{(1)}+k_{12}^{(2)} & 0 & 0\\
k_{21}^{(1)}+k_{21}^{(2)} & k_{22}^{(1)}+k_{22}^{(2)}+k_{11}^{(3)} & k_{12}^{(3)} & 0\\
0 & k_{21}^{(3)} & k_{22}^{(3)}+k_{11}^{(4)}+k_{11}^{(5)} & k_{12}^{(4)}+k_{12}^{(5)}\\
0 & 0 & k_{21}^{(4)}+k_{21}^{(5)} & k_{22}^{(4)}+k_{22}^{(5)}
\end{bmatrix}
\begin{Bmatrix} u_1\\ u_2\\ u_3\\ u_4 \end{Bmatrix}
=
\begin{Bmatrix}
Q_1^{(1)}+Q_1^{(2)}\\
Q_2^{(1)}+Q_2^{(2)}+Q_1^{(3)}\\
Q_2^{(3)}+Q_1^{(4)}+Q_1^{(5)}\\
Q_2^{(4)}+Q_2^{(5)}
\end{Bmatrix}
\tag{2.36}
$$

which is the assembled system.

Table 2.2 Connectivity: global node numbers for 5-spring example problem

	Element 1	Element 2	Element 3	Element 4	Element 5
Local node 1	1	1	2	3	3
Local node 2	2	2	3	4	4

Assembly follows the automated process shown in Fig. 2.6, where contributions of each element are bracketed for clarity. The global system of governing equilibrium equations is

$$
\begin{bmatrix}
K_{11} & K_{12} & K_{13} & K_{14}\\
K_{21} & K_{22} & K_{23} & K_{24}\\
K_{31} & K_{32} & K_{33} & K_{34}\\
K_{41} & K_{42} & K_{43} & K_{44}
\end{bmatrix}
\begin{Bmatrix} u_1\\ u_2\\ u_3\\ u_4 \end{Bmatrix}
=
\begin{Bmatrix} Q_1\\ Q_2\\ Q_3\\ Q_4 \end{Bmatrix}
\leftrightarrow [\boldsymbol{K}]\{\boldsymbol{u}\}=\{\boldsymbol{Q}\}.
\tag{2.37}
$$

Note that the size of $\boldsymbol{K}$ is 4×4 since the connected mesh contains 4 nodes with one degree-of-freedom per node. At this juncture, the system of equations is statically indeterminate. Application of boundary conditions will alleviate the indeterminancy and permit a solution.

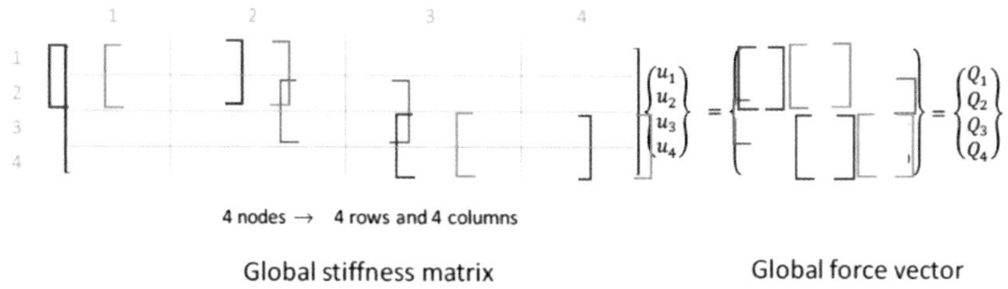

Fig. 2.6 Example problem: assembly of global equilibrium equations

2.2.5 An Example: Boundary Conditions

Recall from discussion in §2.2.1 that statically indeterminant problems are generally under-constrained, with a corresponding stiffness matrix being singular and thus having a null determinant, i.e., $\det \boldsymbol{K} = 0$ in the present context. This is the case for the global stiffness matrix in (2.36).

Recall also that two types of boundary conditions can be applied to render a problem statically determinant:

- **Essential boundary conditions**: displacement type for mechanical problems, also called **Dirichlet boundary conditions**;
- **Natural boundary conditions**: force type for mechanical problems, also called **Neumann boundary conditions.**

Boundary conditions can be applied anywhere in a physical problem, but in FEA they are ultimately implemented at the nodes. At a given node in a mechanical problem, *the absence of an imposed displacement or force is equivalent to an applied net force condition of zero magnitude.*

Returning to the 5-spring example problem, consider the assembled force vector on the right side of (2.36):

$$\begin{Bmatrix} Q_1 \\ Q_2 \\ Q_3 \\ Q_4 \end{Bmatrix} = \begin{Bmatrix} Q_1^{(1)} + Q_1^{(2)} \\ Q_2^{(1)} + Q_2^{(2)} + Q_1^{(3)} \\ Q_2^{(3)} + Q_1^{(4)} + Q_1^{(5)} \\ Q_2^{(4)} + Q_2^{(5)} \end{Bmatrix}. \tag{2.38}$$

Recall from §2.2.3 that the imposed boundary conditions at nodes 1 and 4 are

$$u_1 = g, \qquad Q_4 = P. \tag{2.39}$$

The force acting at node 1, Q_1, is the reaction due to the essential boundary condition imposed there, and its value is unknown until the problem is solved. It is important to note that u_1 and Q_1 cannot both be imposed, i.e., a natural condition and an essential condition cannot be prescribed simultaneously at the same point for the same degree-of-freedom and its work conjugate. Since no explicit conditions are imposed on nodes 2 and 3, net forces at these locations vanish:

$$Q_2 = Q_3 = 0. \tag{2.40}$$

Consider in further detail, for example, the condition on node 2:

$$Q_2 = Q_2^{(1)} + Q_2^{(2)} + Q_1^{(3)} = 0. \tag{2.41}$$

This condition can be interpreted physically in a number of ways:

- Q_2 is the total net force at node 2 due to internal force contributions from elements 1, 2, and 3;
- The forces in elements 1, 2, and 3 are in equilibrium with the net force Q_2 acting at node 2;
- The internal forces in elements 1, 2, and 3 at node 2 support an externally applied force Q_2.

In each interpretation of this specific problem, Q_2 is of zero magnitude, but a similar description would apply if Q_2 were nonzero. In summary, the natural boundary conditions for this 5-spring problem are $Q_2 = Q_3 = 0$ and $Q_4 = P$.

Now we implement the essential boundary condition $u_1 = g$. A straightforward substitution for the first row in (2.37) gives

$$\begin{bmatrix} 1 & 0 & 0 & 0 \\ K_{21} & K_{22} & K_{23} & K_{24} \\ K_{31} & K_{32} & K_{33} & K_{34} \\ K_{41} & K_{42} & K_{43} & K_{44} \end{bmatrix} \begin{Bmatrix} u_1 \\ u_2 \\ u_3 \\ u_4 \end{Bmatrix} = \begin{Bmatrix} g \\ Q_2 \\ Q_3 \\ Q_4 \end{Bmatrix}. \tag{2.42}$$

Next we move all contributions of u_1 to the right side of the equation and thereby eliminate the first row and first column of the global stiffness:

$$\begin{bmatrix} K_{22} & K_{23} & K_{24} \\ K_{32} & K_{33} & K_{34} \\ K_{42} & K_{43} & K_{44} \end{bmatrix} \begin{Bmatrix} u_2 \\ u_3 \\ u_4 \end{Bmatrix} = \begin{Bmatrix} Q_2 \\ Q_3 \\ Q_4 \end{Bmatrix} - \begin{Bmatrix} K_{21} \\ K_{31} \\ K_{41} \end{Bmatrix} g. \tag{2.43}$$

This may be written as

$$[\boldsymbol{K}]\{\boldsymbol{u}\} = \{\boldsymbol{F}\}, \tag{2.44}$$

where $\boldsymbol{F}$ is vector of dimension three containing all terms on the right side of (2.43), and now $\boldsymbol{K}$ is a 3×3 reduced global stiffness matrix, often called the condensed stiffness matrix.

2.2.6 An Example: Solving

For static problems, the basic role of a solver in FEA is inversion of the stiffness matrix, which tends to involve its diagonalization. Equation (2.44), which is general enough to apply to nearly any class of static problem, is solved for the vector-valued primary solution variable $\boldsymbol{u}$:

$$[\boldsymbol{K}]\{\boldsymbol{u}\} = \{\boldsymbol{F}\} \rightarrow [\boldsymbol{K}]^{-1}[\boldsymbol{K}]\{\boldsymbol{u}\} = [\boldsymbol{K}]^{-1}\{\boldsymbol{F}\} \rightarrow \{\boldsymbol{u}\} = [\boldsymbol{K}]^{-1}\{\boldsymbol{F}\}. \tag{2.45}$$

Solution of the problem thus reduces to computation of $\boldsymbol{K}^{-1}$ and then matrix-vector multiplication for the unknowns.

Returning to the 5-spring example, substituting k for the spring constant of all five elements and imposing the boundary conditions $u_1 = g = 0 = Q_2 = Q_3$ and $Q_4 = P$, the reduced global system becomes

$$\begin{bmatrix} 3k & -k & 0 \\ -k & 3k & -2k \\ 0 & -2k & 2k \end{bmatrix} \begin{Bmatrix} u_2 \\ u_3 \\ u_4 \end{Bmatrix} = \begin{Bmatrix} 0 \\ 0 \\ P \end{Bmatrix}. \tag{2.46}$$

Note that the condensed stiffness matrix on the left is symmetric and non-singular, with a determinant of magnitude $|\boldsymbol{K}| = 8k^2$. The solution of (2.46) for the primary variables is

$$u_2 = \frac{P}{2k}, \qquad u_3 = \frac{3P}{2k}, \qquad u_4 = \frac{2P}{k}. \tag{2.47}$$

It is emphasized that we solved the *global* equations for the unknown displacements. The *local* element equations do not directly yield the problem solution because prior to assembly, most individual elements are "unaware" of the boundary conditions that enable existence of a solution to the problem.

2.2.7 An Example: Post-processing

Solution of the global system results in the vector of primary unknowns, the displacements in this case. To complete the analysis, one may wish to investigate the force in each spring, the reaction force at the boundary, and perhaps other information distinct from $\boldsymbol{u}$. Post-processing involves use of the local element equations to determine such secondary information.

As an explicit example, we return to the 5-spring problem and compute the internal force in element 1. Substituting the known values $u_1 = 0$ and $u_2 = P/(2k)$, the former imposed and the latter found from the global solution (2.47), the equilibrium equations for element 1 in (2.33) become

$$\begin{Bmatrix} Q_1^{(1)} \\ Q_2^{(1)} \end{Bmatrix} = \begin{bmatrix} k_{11}^{(1)} & k_{12}^{(1)} \\ k_{21}^{(1)} & k_{22}^{(1)} \end{bmatrix} \begin{Bmatrix} u_1^{(1)} \\ u_2^{(1)} \end{Bmatrix} = \begin{bmatrix} k & -k \\ -k & k \end{bmatrix} \begin{Bmatrix} 0 \\ P/(2k) \end{Bmatrix} = \begin{Bmatrix} -P/2 \\ P/2 \end{Bmatrix}. \tag{2.48}$$

The same procedure can be used for all other elements. Finally, the total reaction force at node 1 can be determined by adding the contributions from elements 1 and 2:

$$Q_1 = Q_1^{(1)} + Q_1^{(2)} = (-P/2) + (-P/2) = -P, \tag{2.49}$$

which, as physically expected, counteracts the total applied force P at the right end of the system in Fig. 2.4.

2.2.8 Direct Method in 2-D

Two- and three-dimensional structures can be constructed from 1-D members, i.e., members such as bars and springs that can support only axial forces. The nodes of such structures can move in 2-D or 3-D space. The present discussion pertains to 2-D space trusses; 3-D trusses are discussed later in §2.2.9.

For 1-D trusses, only a single spatial coordinate (e.g., x) is necessary, and all local and global element equations are referred to the same (trivial 1-D) coordinate system aligned parallel to the direction of forces and displacements. In contrast, for 2-D trusses, a 2-D global coordinate system is necessary, as is an additional step transforming the local equilibrium equations referred to each truss's axis to equations referred to the global coordinate system.

Refer to Fig. 2.7, which introduces local and global coordinate systems. The local coordinate system for an element consists of the longitudinal and transverse directions denoted by respective subscripts L and T. Global directions are denoted by Cartesian coordinates x and y. The local coordinate frame is obtained from the global frame via rotation of the latter counterclockwise by angle θ. This angle varies from element to element, but we suppress the e superscript for now. Displacements in the two coordinate systems are related by the vector-matrix expression

$$\begin{Bmatrix} u_{1L} \\ u_{1T} \\ u_{2L} \\ u_{2T} \end{Bmatrix} = \begin{bmatrix} \cos\theta & \sin\theta & 0 & 0 \\ -\sin\theta & \cos\theta & 0 & 0 \\ 0 & 0 & \cos\theta & \sin\theta \\ 0 & 0 & -\sin\theta & \cos\theta \end{bmatrix} \begin{Bmatrix} u_{1x} \\ u_{1y} \\ u_{2x} \\ u_{2y} \end{Bmatrix} \leftrightarrow \{\boldsymbol{u}_{LT}^{(e)}\} = [\boldsymbol{T}^{(e)}]\{\boldsymbol{u}_{xy}^{(e)}\}. \tag{2.50}$$

The transformation matrix is orthogonal, meaning its inverse ($[\cdot]^{-1}$) and transpose ($[\cdot]^{\mathrm{T}}$) are equal. Using the short-hand notation

$$c\theta^{(e)} = \cos\theta, \qquad s\theta^{(e)} = \sin\theta, \tag{2.51}$$

the transformation matrix and its inverse are respectively

$$[\boldsymbol{T}^{(e)}] = \begin{bmatrix} c\theta^{(e)} & s\theta^{(e)} & 0 & 0 \\ -s\theta^{(e)} & c\theta^{(e)} & 0 & 0 \\ 0 & 0 & c\theta^{(e)} & s\theta^{(e)} \\ 0 & 0 & -s\theta^{(e)} & c\theta^{(e)} \end{bmatrix}, \tag{2.52}$$

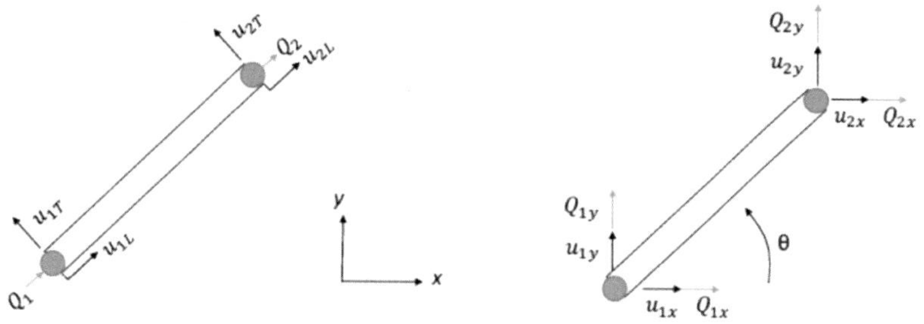

Fig. 2.7 Truss element quantities referred to local (left) and global (right) coordinate systems

$$[\boldsymbol{T}^{(e)}]^{-1} = [\boldsymbol{T}^{(e)}]^{\mathrm{T}} = \begin{bmatrix} c\theta^{(e)} & -s\theta^{(e)} & 0 & 0 \\ s\theta^{(e)} & c\theta^{(e)} & 0 & 0 \\ 0 & 0 & c\theta^{(e)} & -s\theta^{(e)} \\ 0 & 0 & s\theta^{(e)} & c\theta^{(e)} \end{bmatrix}. \tag{2.53}$$

Displacements referred to the global system can be obtained using the latter, and the same transformation laws apply for the force vectors:

$$\{\boldsymbol{u}_{xy}^{(e)}\} = [\boldsymbol{T}^{(e)}]^{\mathrm{T}}\{\boldsymbol{u}_{LT}^{(e)}\}, \qquad \{\boldsymbol{Q}_{xy}^{(e)}\} = [\boldsymbol{T}^{(e)}]^{\mathrm{T}}\{\boldsymbol{Q}_{LT}^{(e)}\}; \tag{2.54}$$

$$\{\boldsymbol{u}_{LT}^{(e)}\} = [\boldsymbol{T}^{(e)}]\{\boldsymbol{u}_{xy}^{(e)}\}, \qquad \{\boldsymbol{Q}_{LT}^{(e)}\} = [\boldsymbol{T}^{(e)}]\{\boldsymbol{Q}_{xy}^{(e)}\}. \tag{2.55}$$

It is remarked that transverse components of $\boldsymbol{Q}_{LT}^{(e)}$ are zero by construction, since trusses cannot support transverse forces.

Now consider the element equilibrium equation. Referred to the global 2-D coordinate system, static equilibrium for a single truss, bar, or spring dictates

$$[\boldsymbol{k}_{xy}^{(e)}]\{\boldsymbol{u}_{xy}^{(e)}\} = \{\boldsymbol{Q}_{xy}^{(e)}\}. \tag{2.56}$$

Similarly, referred to the local coordinate system for that element,

$$[\boldsymbol{k}_{LT}^{(e)}]\{\boldsymbol{u}_{LT}^{(e)}\} = \{\boldsymbol{Q}_{LT}^{(e)}\}. \tag{2.57}$$

Substituting (2.55) into (2.57) and then invoking (2.53),

$$[\boldsymbol{k}_{LT}^{(e)}][\boldsymbol{T}^{(e)}]\{\boldsymbol{u}_{xy}^{(e)}\} = [\boldsymbol{T}^{(e)}]\{\boldsymbol{Q}_{xy}^{(e)}\} \Rightarrow [\boldsymbol{T}^{(e)}]^{\mathrm{T}}[\boldsymbol{k}_{LT}^{(e)}][\boldsymbol{T}^{(e)}]\{\boldsymbol{u}_{xy}^{(e)}\} = \{\boldsymbol{Q}_{xy}^{(e)}\}. \tag{2.58}$$

This equality can be rewritten as in (2.56), leading to the transformation rule for the stiffness matrix:

$$[\boldsymbol{k}_{xy}^{(e)}] = [\boldsymbol{T}^{(e)}]^{\mathrm{T}}[\boldsymbol{k}_{LT}^{(e)}][\boldsymbol{T}^{(e)}]. \tag{2.59}$$

The stiffness matrix for a single element in its local system corresponds to an element aligned with the x-axis. Since transverse forces vanish but transverse displacements do not, in general, the local element stiffness for a 2-D truss can be obtained by expanding the 2×2 element stiffness of (2.25) to a 4×4 matrix by inserting "null" rows and columns 2 and 4:

$$[\boldsymbol{k}_{LT}^{(e)}] = k \begin{bmatrix} 1 & 0 & -1 & 0 \\ 0 & 0 & 0 & 0 \\ -1 & 0 & 1 & 0 \\ 0 & 0 & 0 & 0 \end{bmatrix}. \tag{2.60}$$

Recall also that for a simple bar, $k = EA/L$ as in (2.19). Invoking (2.59), the element stiffness referred to global coordinates becomes

$$\begin{aligned} [\boldsymbol{k}_{xy}^{(e)}] &= [\boldsymbol{T}^{(e)}]^{\mathrm{T}} k \begin{bmatrix} 1 & 0 & -1 & 0 \\ 0 & 0 & 0 & 0 \\ -1 & 0 & 1 & 0 \\ 0 & 0 & 0 & 0 \end{bmatrix} [\boldsymbol{T}^{(e)}] \\ &= k \begin{bmatrix} (c\theta^{(e)})^2 & c\theta^{(e)}s\theta^{(e)} & -(c\theta^{(e)})^2 & -c\theta^{(e)}s\theta^{(e)} \\ c\theta^{(e)}s\theta^{(e)} & (s\theta^{(e)})^2 & -c\theta^{(e)}s\theta^{(e)} & -(s\theta^{(e)})^2 \\ -(c\theta^{(e)})^2 & -c\theta^{(e)}s\theta^{(e)} & (c\theta^{(e)})^2 & c\theta^{(e)}s\theta^{(e)} \\ -c\theta^{(e)}s\theta^{(e)} & -(s\theta^{(e)})^2 & c\theta^{(e)}s\theta^{(e)} & (s\theta^{(e)})^2 \end{bmatrix}. \end{aligned} \tag{2.61}$$

In global coordinates, the equilibrium equations in vector-matrix form for a single truss element are

$$k \begin{bmatrix} (c\theta^{(e)})^2 & c\theta^{(e)}s\theta^{(e)} & -(c\theta^{(e)})^2 & -c\theta^{(e)}s\theta^{(e)} \\ c\theta^{(e)}s\theta^{(e)} & (s\theta^{(e)})^2 & -c\theta^{(e)}s\theta^{(e)} & -(s\theta^{(e)})^2 \\ -(c\theta^{(e)})^2 & -c\theta^{(e)}s\theta^{(e)} & (c\theta^{(e)})^2 & c\theta^{(e)}s\theta^{(e)} \\ -c\theta^{(e)}s\theta^{(e)} & -(s\theta^{(e)})^2 & c\theta^{(e)}s\theta^{(e)} & (s\theta^{(e)})^2 \end{bmatrix} \begin{Bmatrix} u_{1x}^{(e)} \\ u_{1y}^{(e)} \\ u_{2x}^{(e)} \\ u_{2y}^{(e)} \end{Bmatrix} = \begin{Bmatrix} Q_{1x}^{(e)} \\ Q_{1y}^{(e)} \\ Q_{2x}^{(e)} \\ Q_{2y}^{(e)} \end{Bmatrix}. \tag{2.62}$$

In 1-D, a single truss element has two equations, one per each nodal degree-of-freedom. In 2-D, a single truss element has four equations since each node has two degrees-of-freedom, i.e., displacement components in x- and y-directions.

Another example problem will now be used to demonstrate steps in the DM for 2-D trusses: the problem setup, assembly, boundary conditions, solution, and post-processing. This example is shown in Fig. 2.8. Two linear trusses are oriented perpendicular to one another, with an external force applied at their intersection and directed diagonally rightwards and downwards.

Examining Fig. 2.8, element 1 contains global nodes 1 and 2, while element 2 contains global nodes 2 and 3. Since global node 2 is shared, we have the constraints

$$u_{2L}^{(1)} = -u_{1T}^{(2)} = u_{2x}, \qquad u_{2T}^{(1)} = u_{1L}^{(2)} = u_{2y}. \tag{2.63}$$

Orientations of the elements lead to

$$\theta^{(1)} = 0, \quad \theta^{(2)} = 90^\circ \Rightarrow c\theta^{(1)} = s\theta^{(2)} = 1, \quad c\theta^{(2)} = s\theta^{(1)} = 0. \tag{2.64}$$

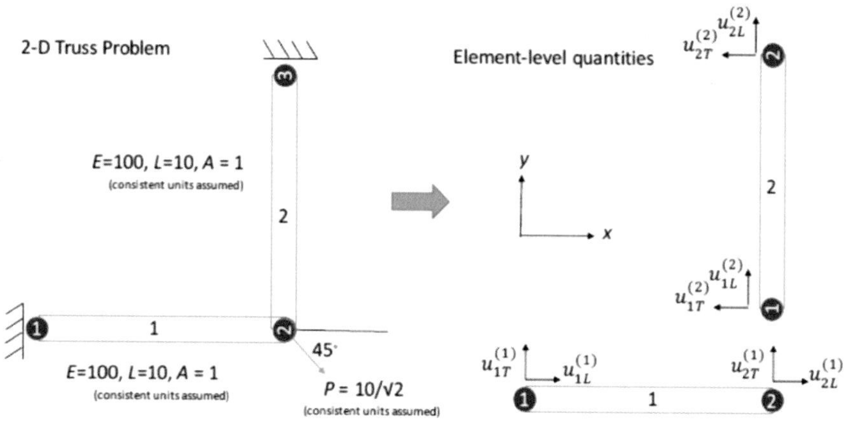

Fig. 2.8 2-D truss problem (left) decomposed into two finite elements (right)

Element stiffness components are of the following magnitudes:

$$[k^{(1)} = (EA/L)^{(1)} = 10, \quad k^{(2)} = (EA/L)^{(2)} = 10] \Rightarrow k^{(1)} = k^{(2)} = k = 10. \tag{2.65}$$

Substitution of (2.63), (2.64), and (2.65) into the 2-D element equilibrium equations of (2.62) gives results for respective elements 1 and 2:

$$k\begin{bmatrix} 1 & 0 & -1 & 0 \\ 0 & 0 & 0 & 0 \\ -1 & 0 & 1 & 0 \\ 0 & 0 & 0 & 0 \end{bmatrix}\begin{Bmatrix} u_{1x} \\ u_{1y} \\ u_{2x} \\ u_{2y} \end{Bmatrix} = \begin{Bmatrix} Q_{1x}^{(1)} \\ Q_{1y}^{(1)} \\ Q_{2x}^{(1)} \\ Q_{2y}^{(1)} \end{Bmatrix}, \tag{2.66}$$

$$k\begin{bmatrix} 0 & 0 & 0 & 0 \\ 0 & 1 & 0 & -1 \\ 0 & 0 & 0 & 0 \\ 0 & -1 & 0 & 1 \end{bmatrix}\begin{Bmatrix} u_{2x} \\ u_{2y} \\ u_{3x} \\ u_{3y} \end{Bmatrix} = \begin{Bmatrix} Q_{1x}^{(2)} \\ Q_{1y}^{(2)} \\ Q_{2x}^{(2)} \\ Q_{2y}^{(2)} \end{Bmatrix}. \tag{2.67}$$

Assembling the two sets of equations leads to the following global system of static equilibrium:

$$k\begin{bmatrix} 1 & 0 & -1 & 0 & 0 & 0 \\ 0 & 0 & 0 & 0 & 0 & 0 \\ -1 & 0 & 1 & 0 & 0 & 0 \\ 0 & 0 & 0 & 1 & 0 & -1 \\ 0 & 0 & 0 & 0 & 0 & 0 \\ 0 & 0 & 0 & -1 & 0 & 1 \end{bmatrix}\begin{Bmatrix} u_{1x} \\ u_{1y} \\ u_{2x} \\ u_{2y} \\ u_{3x} \\ u_{3y} \end{Bmatrix} = \begin{Bmatrix} Q_{1x}^{(1)} \\ Q_{1y}^{(1)} \\ Q_{2x}^{(1)} + Q_{1x}^{(2)} \\ Q_{2y}^{(1)} + Q_{1y}^{(2)} \\ Q_{2x}^{(2)} \\ Q_{2y}^{(2)} \end{Bmatrix} = \begin{Bmatrix} Q_{1x} \\ Q_{1y} \\ Q_{2x} \\ Q_{2y} \\ Q_{3x} \\ Q_{3y} \end{Bmatrix}. \tag{2.68}$$

With assembly complete, boundary conditions are now applied. Referring again to Fig. 2.8, the essential boundary conditions correspond to fixed coordinates (null displacements) of global nodes 1 and 3:

$$u_{1x} = u_{1y} = 0, \qquad u_{3x} = u_{3y} = 0. \tag{2.69}$$

The natural boundary conditions are associated with the applied force P at global node 2:

$$Q_{2x} = P\cos 45^\circ = 5, \qquad Q_{2y} = -P\sin 45^\circ = -5. \tag{2.70}$$

Invoking the four essential boundary conditions of (2.69) reduces the global system in (2.68) to two equations in two unknowns u_{2x} and u_{2y}:

$$k\begin{bmatrix} 1 & 0 \\ 0 & 1 \end{bmatrix}\begin{Bmatrix} u_{2x} \\ u_{2y} \end{Bmatrix} = \begin{Bmatrix} Q_{2x} \\ Q_{2y} \end{Bmatrix} = \frac{\sqrt{2}P}{2}\begin{Bmatrix} 1 \\ -1 \end{Bmatrix}, \tag{2.71}$$

where the natural boundary conditions have been symbolically substituted on the right side. Notice that rows and columns corresponding to imposed null displacements for nodes 1 and 3 have been eliminated, reducing the global stiffness matrix from size 6×6 to its condensed version that is 2×2. Inversion of the stiffness matrix on the left side of (2.71) is trivial. The primary solution follows as

$$\begin{Bmatrix} u_{2x} \\ u_{2y} \end{Bmatrix} = \frac{\sqrt{2}P}{2k}\begin{Bmatrix} 1 \\ -1 \end{Bmatrix} = \begin{Bmatrix} 0.5 \\ -0.5 \end{Bmatrix}. \tag{2.72}$$

As an example of post-processing, we now use the primary solution to compute stresses in each element. The element constitutive equation relating stress σ to strain ε is, for a local coordinate system aligned parallel to the loading direction,

$$\sigma^{(e)} = E^{(e)}\varepsilon^{(e)} = E^{(e)}\frac{\mathrm{d}u^{(e)}}{\mathrm{d}x} = E^{(e)}\frac{u_{2L}^{(e)} - u_{1L}^{(e)}}{L^{(e)}}. \tag{2.73}$$

This equation assumes that strain, and therefore stress, are constant within an element. We now transform this equation to global coordinates since the primary solution is referred to the global 2-D frame. Writing (2.73) in vector-matrix form and invoking (2.55) gives

$$\sigma^{(e)} = \frac{E^{(e)}}{L^{(e)}}[-1 \quad 1]\begin{Bmatrix} u_{1L}^{(e)} \\ u_{2L}^{(e)} \end{Bmatrix} = \frac{E^{(e)}}{L^{(e)}}[-1 \quad 0 \quad 1 \quad 0]\begin{Bmatrix} u_{1L}^{(e)} \\ u_{1T}^{(e)} \\ u_{2L}^{(e)} \\ u_{2T}^{(e)} \end{Bmatrix}$$

$$= \frac{E^{(e)}}{L^{(e)}}[-1 \quad 0 \quad 1 \quad 0][\boldsymbol{T}^{(e)}]\begin{Bmatrix} u_{1x}^{(e)} \\ u_{1y}^{(e)} \\ u_{2x}^{(e)} \\ u_{2y}^{(e)} \end{Bmatrix}, \tag{2.74}$$

where the coordinate transformation matrix $\boldsymbol{T}^{(e)}$ is defined in (2.52) in terms of orientation $\theta^{(e)}$ of the element. Substituting the solution in (2.72) as well as the essential boundary conditions in (2.69) into this expression, where $e = 1, 2$, gives the element stresses:

$$\sigma^{(1)} = \frac{E^{(1)}}{L^{(1)}}[-1 \quad 0 \quad 1 \quad 0]\begin{bmatrix} 1 & 0 & 0 & 0 \\ 0 & 1 & 0 & 0 \\ 0 & 0 & 1 & 0 \\ 0 & 0 & 0 & 1 \end{bmatrix}\begin{Bmatrix} 0 \\ 0 \\ 0.5 \\ -0.5 \end{Bmatrix} = \frac{E^{(1)}}{2L^{(1)}} = 5, \tag{2.75}$$

$$\sigma^{(2)} = \frac{E^{(2)}}{L^{(2)}}[-1 \quad 0 \quad 1 \quad 0]\begin{bmatrix} 0 & 1 & 0 & 0 \\ -1 & 0 & 0 & 0 \\ 0 & 0 & 0 & 1 \\ 0 & 0 & -1 & 0 \end{bmatrix}\begin{Bmatrix} 0.5 \\ -0.5 \\ 0 \\ 0 \end{Bmatrix} = \frac{E^{(2)}}{2L^{(2)}} = 5. \tag{2.76}$$

This result is physically expected from the symmetry of the structure with respect to the loading: tensile stresses of equal magnitude are supported by each finite element or member of the truss.

2.2.9 Direct Method in 3-D

Now we consider truss structures in three spatial dimensions, i.e., 3-D space trusses. Analogously to the 2-D case, 3-D trusses are merely coordinate transformations and subsequent assembly of 1-D trusses. The procedure of analysis is the same as that for 2-D trusses except for the added complexity of a 3-D coordinate transformation law and bookkeeping of three degrees-of-freedom at each node corresponding to displacements in x-, y-, and z-directions.

Transformation formulae are designated as follows upon consideration of the coordinate system and notation in Fig. 2.9. We consider a single truss (i.e., bar or spring) element whose local equilibrium equations are of the usual form

$$\begin{bmatrix} k_{11}^{(e)} & k_{12}^{(e)} \\ k_{21}^{(e)} & k_{22}^{(e)} \end{bmatrix}\begin{Bmatrix} u_1^{(e)} \\ u_2^{(e)} \end{Bmatrix} = \begin{Bmatrix} Q_1^{(e)} \\ Q_2^{(e)} \end{Bmatrix} \leftrightarrow [\boldsymbol{k}^{(e)}]\{\boldsymbol{u}^{(e)}\} = \{\boldsymbol{Q}^{(e)}\}. \tag{2.77}$$

This set of equations can be transformed to a representation in the global xyz coordinate system of the form

$$[\boldsymbol{k}_{xyz}^{(e)}]\{\boldsymbol{u}_{xyz}^{(e)}\} = \{\boldsymbol{Q}_{xyz}^{(e)}\}, \tag{2.78}$$

where via use of a 2×6 orthogonal transformation matrix $\boldsymbol{T}^{(e)}$,

$$\{\boldsymbol{u}^e\} = [\boldsymbol{T}^{(e)}]\{\boldsymbol{u}_{xyz}^{(e)}\}, \qquad \{\boldsymbol{Q}^{(e)}\} = [\boldsymbol{T}^{(e)}]\{\boldsymbol{Q}_{xyz}^{(e)}\}, \tag{2.79}$$

with global displacement and force vectors having six components each. The transformation rule for the element stiffness matrix is also analogous to that for the 2-D case:

$$[\boldsymbol{k}_{xyz}^{(e)}] = [\boldsymbol{T}^{(e)}]^{\mathrm{T}}[\boldsymbol{k}^{(e)}][\boldsymbol{T}^{(e)}], \tag{2.80}$$

where $\boldsymbol{k}_{xyz}^{(e)}$ is a 6×6 matrix.

Referring to Fig. 2.9, entries of the transformation matrix for 3-D trusses are obtained as follows. Node 1 is located at (x_1, y_1, z_1), and node 2 is located at (x_2, y_2, z_2). The length of the element is simply

$$L = \sqrt{(x_2 - x_1)^2 + (y_2 - y_1)^2 + (z_2 - z_1)^2}. \tag{2.81}$$

Direction cosines that describe the orientation of the element e are defined as the following ratios:

$$l^{(e)} = \frac{x_2 - x_1}{L}, \qquad m^{(e)} = \frac{y_2 - y_1}{L}, \qquad n^{(e)} = \frac{z_2 - z_1}{L}. \tag{2.82}$$

The transformation matrix is then defined, here without proof, as

$$[\boldsymbol{T}^{(e)}] = \begin{bmatrix} l^{(e)} & m^{(e)} & n^{(e)} & 0 & 0 & 0 \\ 0 & 0 & 0 & l^{(e)} & m^{(e)} & n^{(e)} \end{bmatrix}. \tag{2.83}$$

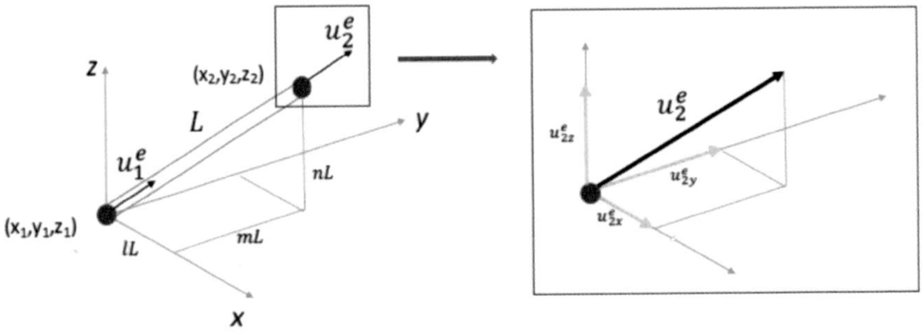

Fig. 2.9 3-D truss element (left) and a nodal displacement vector decomposed into orthogonal components (right)

Temporarily suppressing the (e) superscript on direction cosines, etc., and writing in terms of element stiffness constant $k = EA/L$, the element stiffness matrix in global coordinates in (2.80) becomes

$$[\boldsymbol{k}_{xyz}^{(e)}] \rightarrow [\boldsymbol{k}_{xyz}] = k \begin{bmatrix} l^2 & lm & ln & -l^2 & -lm & -ln \\ lm & m^2 & mn & -lm & -m^2 & -mn \\ ln & mn & n^2 & -ln & -mn & n^2 \\ -l^2 & -lm & -ln & l^2 & lm & ln \\ -lm & -m^2 & -mn & lm & m^2 & mn \\ -ln & -mn & -n^2 & ln & mn & n^2 \end{bmatrix}, \tag{2.84}$$

which is notably symmetric.

Once the problem is defined and transformed to global 3-D Cartesian coordinates, the assembly and solution procedures are identical to those for 2-D trusses. Post-processing follows analogously, though for some quantities one must logically consider 3-D transformation rules. For example, the stress in a truss element within a 3-D structure is computed from nodal displacements $\boldsymbol{u}_{xyz}^{(e)}$ as follows:

$$\begin{aligned} \sigma^{(e)} &= \frac{E^{(e)}}{L^{(e)}}[-1 \quad 1]\begin{Bmatrix} u_1^{(e)} \\ u_2^{(e)} \end{Bmatrix} = \frac{E^{(e)}}{L^{(e)}}[-1 \quad 1][\boldsymbol{T}^{(e)}]\begin{Bmatrix} u_{1x}^{(e)} \\ u_{1y}^{(e)} \\ u_{1z}^{(e)} \\ u_{2x}^{(e)} \\ u_{2y}^{(e)} \\ u_{2z}^{(e)} \end{Bmatrix} \\ &= \frac{E^{(e)}}{L^{(e)}}[-l^{(e)} \quad -m^{(e)} \quad -n^{(e)} \quad l^{(e)} \quad m^{(e)} \quad n^{(e)}]\begin{Bmatrix} u_{1x}^{(e)} \\ u_{1y}^{(e)} \\ u_{1z}^{(e)} \\ u_{2x}^{(e)} \\ u_{2y}^{(e)} \\ u_{2z}^{(e)} \end{Bmatrix}. \end{aligned} \tag{2.85}$$

2.2.10 Incorporation of Thermal Strain

Changes in temperature will cause a stress-free elastic body to expand or contract. In the 1-D case, the total strain ε can be separated into a mechanically elastic part ε_e associated with the stress σ and a thermal part ε_t that does not contribute directly to stress:

$$\varepsilon = \varepsilon_e + \varepsilon_t; \qquad \varepsilon_e = \sigma/E, \quad \varepsilon_t = \alpha\Delta T. \tag{2.86}$$

Temperature is T, and temperature change from a reference state is denoted by ΔT. The coefficient of thermal expansion is α, a positive scalar for nearly all materials, leading to expansion with temperature rise. The force acting on one node of a bar or truss element with cross-sectional area A is derived by inspection in the Direct Method as

$$\begin{aligned} Q_1 &= A\sigma = EA\varepsilon_e = EA(\varepsilon - \varepsilon_t) \\ &= EA\left(\frac{\mathrm{d}u}{\mathrm{d}x} - \alpha\Delta T\right) = \frac{EA}{L}(u_1 - u_2) - EA\alpha\Delta T. \end{aligned} \tag{2.87}$$

The same procedure may be used to evaluate Q_2, the force acting on the other node (see Fig. 2.2). Noting that the contribution of temperature change does not contain displacements, this contribution manifests as an effective force. The equilibrium equation for a single element e (see §2.2.1) is thus

$$\begin{bmatrix} k_{11}^{(e)} & k_{12}^{(e)} \\ k_{21}^{(e)} & k_{22}^{(e)} \end{bmatrix} \begin{Bmatrix} u_1^{(e)} \\ u_2^{(e)} \end{Bmatrix} = \begin{Bmatrix} Q_1^{(e)} \\ Q_2^{(e)} \end{Bmatrix} + E^{(e)}A^{(e)}\alpha^{(e)}\Delta T \begin{Bmatrix} -1 \\ 1 \end{Bmatrix}, \tag{2.88}$$

where thermoelastic properties and cross-sectional areas may vary from element to element. Following similar derivation steps for 2-D and 3-D trusses, the right-hand sides of element equilibrium equations in §2.2.8 and §2.2.9 are augmented as follows:

$$\begin{Bmatrix} Q_{1x}^{(e)} \\ Q_{1y}^{(e)} \\ Q_{2x}^{(e)} \\ Q_{2y}^{(e)} \end{Bmatrix} \to \begin{Bmatrix} Q_{1x}^{(e)} \\ Q_{1y}^{(e)} \\ Q_{2x}^{(e)} \\ Q_{2y}^{(e)} \end{Bmatrix} + E^{(e)}A^{(e)}\alpha^{(e)}\Delta T \begin{Bmatrix} -c\theta^{(e)} \\ -s\theta^{(e)} \\ c\theta^{(e)} \\ s\theta^{(e)} \end{Bmatrix}, \quad \text{(2-D)}; \tag{2.89}$$

$$\begin{Bmatrix} Q_{1x}^{(e)} \\ Q_{1y}^{(e)} \\ Q_{1z}^{(e)} \\ Q_{2x}^{(e)} \\ Q_{2y}^{(e)} \\ Q_{2z}^{(e)} \end{Bmatrix} \to \begin{Bmatrix} Q_{1x}^{(e)} \\ Q_{1y}^{(e)} \\ Q_{1z}^{(e)} \\ Q_{2x}^{(e)} \\ Q_{2y}^{(e)} \\ Q_{2z}^{(e)} \end{Bmatrix} + E^{(e)}A^{(e)}\alpha^{(e)}\Delta T \begin{Bmatrix} -l^{(e)} \\ -m^{(e)} \\ -n^{(e)} \\ l^{(e)} \\ m^{(e)} \\ n^{(e)} \end{Bmatrix}, \quad \text{(3-D)}. \tag{2.90}$$

It is remarked that the governing equations of thermoelasticity need not be derived via an explicit partitioning of strain into mechanical and thermal parts, though this is the usual approach taken here and in most undergraduate strength-of-materials textbooks. In fact, a more rigorous approach [2] involves use of only a single strain definition but introduction of a Helmholtz free energy function that includes terms associated with elastic strain energy and thermoelastic coupling. For the case of linear thermoelasticity, the governing equilibrium equations are ultimately identical for either theoretical representation, but the same cannot be said for nonlinear thermoelasticity wherein finite strains may be incurred.

2.3 The Finite Element Method

The remainder of Chapter 2 develops the Finite Element Method (FEM) for 1-D structural members. As was the case for the Direct Method (DM) in most of §2.2, we focus attention on fully 1-D problems, where all elements are aligned parallel to a single direction of axial loading. Extension to 2-D and 3-D space trusses follows a procedure identical to DM. Recall from Table 2.1 that FEM and DM are identical except for derivation of the local element equations. Therefore, regardless of which method is used to determine the local equations, the same transformation formulae from local to global coordinate frames apply for consideration of 2-D and 3-D structures. Prior to presentation of the FEM, a brief review of essential operations from differential and integral calculus is in order.

2.3.1 Differential and Integral Calculus

The following mathematical identities, which can be verified in any undergraduate calculus textbook, will be used often later:

- Integration by parts

$$\int_{\Omega} u\mathrm{d}v = (uv)|_{\partial\Omega} - \int_{\Omega} v\mathrm{d}u, \tag{2.91}$$

  where u and v are differentiable functions in domain Ω with boundary $\partial\Omega$;
- Chain rule for differentiation

$$\frac{\mathrm{d}f[y(x)]}{\mathrm{d}x} = \frac{\mathrm{d}f}{\mathrm{d}y}\frac{\mathrm{d}y}{\mathrm{d}x}, \qquad \frac{\partial g[p(x,y),q(x,y)]}{\partial x} = \frac{\partial g}{\partial p}\frac{\partial p}{\partial x} + \frac{\partial g}{\partial q}\frac{\partial q}{\partial x}, \tag{2.92}$$

  where f, y, g, p, and q are differentiable functions of their indicated arguments on the left sides of each equality;
- Product rule for differentiation

$$f(x) = g(x)h(x) \quad \Rightarrow \quad \frac{\mathrm{d}f}{\mathrm{d}x} = \frac{\mathrm{d}g}{\mathrm{d}x}h + \frac{\mathrm{d}h}{\mathrm{d}x}g; \tag{2.93}$$

- Partial differentiation

$$f(x,y) = z(x)x + w(y)y \quad \Rightarrow \quad \frac{\partial f}{\partial x} = \frac{\mathrm{d}z}{\mathrm{d}x}x + z, \quad \frac{\partial f}{\partial y} = \frac{\mathrm{d}w}{\mathrm{d}y}y + w; \tag{2.94}$$

- Associative properties;
- Commutative properties.

Regarding associativity and commutativity, too many possibilities exist to enable inclusion of sufficient examples here, but one should keep in mind that ordinary differentiation is a linear operation.

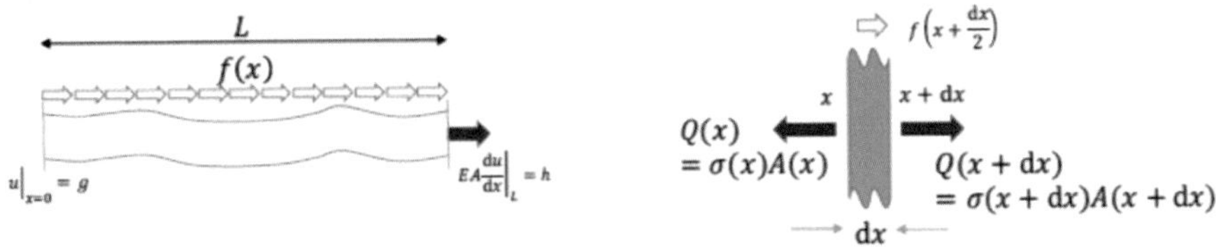

Fig. 2.10 1-D boundary value problem: elastic bar (left) and differential element (right)

2.3.2 Governing Equations: Strong and Weak Forms

In contrast to DM, which invokes heuristic arguments to obtain the governing element equations in algebraic form, FEM begins with a set of one or more ordinary (or partial, depending on the class of problem) differential equation(s). Tools of integral calculus are invoked that transform the so-called strong form of these equation(s) into their so-called weak form. The domain is then discretized into elements, with interpolation functions introduced for field variables at locations distinct from nodal positions. Substitution of interpolation functions into the weak form then ultimately produces the element equilibrium equations in vector-matrix notation.

We begin by considering a 1-D elastic bar or truss. Properties of the bar here may vary with position x: cross-sectional area $A(x)$ and Young's modulus $E(x)$. Also introduced are the displacement function $u(x)$, internal or resultant force $Q(x)$, strain $\varepsilon(x) = \mathrm{d}u/\mathrm{d}x$[1], and stress $\sigma(x) = E(x)\varepsilon(x)$. A body force per unit length, i.e., a distributed load $f(x)$, is also enabled.

The differential equation for static equilibrium at a point x is derived as follows. We address the general 1-D problem shown on the left side of Fig. 2.10, where the essential boundary condition at the left end ($x = 0$) and the natural boundary condition at the right end ($x = L$) are, respectively,

$$u(0) = g, \qquad Q(L) = \sigma(L)A(L) = E(L)A(L)\varepsilon(L) = h. \tag{2.95}$$

Note simply that force = stress × area. A free body diagram of a differential element of the bar of length dx is shown on the right side of Fig. 2.10. Static equilibrium demands that the following sum of forces holds:

$$\sigma(x+\mathrm{d}x)A(x+\mathrm{d}x) - \sigma(x)A(x) + f(x+\mathrm{d}x/2)\mathrm{d}x = 0. \tag{2.96}$$

Divide this equation by dx, then take the limit d$x \to 0$, producing

$$\frac{\mathrm{d}}{\mathrm{d}x}[\sigma(x)A(x)] + f(x) = 0. \tag{2.97}$$

Finally, substitute the constitutive rule $\sigma = E\varepsilon = E(\mathrm{d}u/\mathrm{d}x)$, resulting in

[1] This relationship between displacement and strain can be interpreted as a 1-D compatibility condition.

$$\frac{\mathrm{d}}{\mathrm{d}x}\left[E(x)A(x)\frac{\mathrm{d}u(x)}{\mathrm{d}x}\right]+f(x)=0. \tag{2.98}$$

The strong form of the problem statement is the following. Find $u(x)\forall x\in[0,L]$, given $A(x)$, $E(x)$, and $f(x)$, such that (2.98) holds subject to the boundary conditions in (2.95).

It is emphasized that the strong form is a *complete* statement of a boundary value problem that includes the following information:

- Identification of the (primary) field variable(s) whose value(s) are sought;
- Coordinate system(s) and dimensions of the ambient space (1-D, 2-D, or 3-D);
- The geometry of the body, i.e., the domain and its boundary;
- Requisite properties or parameters of the body;
- The governing differential equation(s);
- Any distributed loads applied within the domain;
- Essential and/or natural boundary conditions.

Consider now a specific example for the 1-D problem. In consistent units, let the bar be tapered with no distributed load such that

$$E(x)A(x)=2L-x, \qquad f(x)=0. \tag{2.99}$$

Substituting (2.99) into (2.98) and integrating once gives, with c an integration constant,

$$\frac{\mathrm{d}}{\mathrm{d}x}\left[(2L-x)\frac{\mathrm{d}u(x)}{\mathrm{d}x}\right]=0\Rightarrow\frac{\mathrm{d}u(x)}{\mathrm{d}x}=\frac{c}{2L-x}. \tag{2.100}$$

The general solution is, with d another constant,

$$u(x)=-c\ln(2L-x)+d. \tag{2.101}$$

Applying boundary conditions, the integration constants are obtained, leading to the particular solution

$$u(x)=-h\ln(2L-x)+g+h\ln(2L). \tag{2.102}$$

In this case, the simple nature of the problem enables an exact analytical solution. In general, the primary field variable $u(x)$ that satisfies the strong form is the exact solution. Numerical techniques benefit from availability of exact solutions because such solutions enable verification and validation.

- **Verification** refers to correct numerical solution of the governing mathematics;

- **Validation** refers to a physically correct solution to a real-world problem.

Once verified and validated, numerical methods can then be used with improved confidence to solve problems for which exact and/or analytical solutions are impractical or impossible to obtain.

The next step involved in derivation of the FEM equations is referred to as weakening the strong form. Returning to the problem shown in Fig. 2.10, we introduce a differentiable scalar function $w(x)$ satisfying $w(0) = 0$ [2]. This function, which in later chapters may be of vector form rather than a scalar, is referred to throughout this textbook as a weight function. It is also sometimes called a test function in the FEA literature.

We first multiply (2.98) by w, giving

$$w\frac{\mathrm{d}}{\mathrm{d}x}\left(EA\frac{\mathrm{d}u}{\mathrm{d}x}\right) + wf = 0. \tag{2.103}$$

Now we integrate this differential equation over the domain $\{\Omega : x \in [0, L]\}$:

$$\int_0^L \left[w\frac{\mathrm{d}}{\mathrm{d}x}\left(EA\frac{\mathrm{d}u}{\mathrm{d}x}\right) + wf\right] \mathrm{d}x = 0. \tag{2.104}$$

Applying integration by parts:

$$\left(wEA\frac{\mathrm{d}u}{\mathrm{d}x}\right)\Bigg|_{x=0}^{x=L} - \int_0^L \left(EA\frac{\mathrm{d}w}{\mathrm{d}x}\frac{\mathrm{d}u}{\mathrm{d}x}\right)\mathrm{d}x + \int_0^L wf\mathrm{d}x = 0. \tag{2.105}$$

Using the assumed condition $w(0) = 0$ and the natural boundary condition in the second of (2.95), we arrive at

$$hw(L) - \int_0^L \left(EA\frac{\mathrm{d}w}{\mathrm{d}x}\frac{\mathrm{d}u}{\mathrm{d}x}\right)\mathrm{d}x + \int_0^L wf\mathrm{d}x = 0. \tag{2.106}$$

The weak form of the problem statement is then the following. Find $u(x) \forall x \in [0, L]$, given $A(x)$, $E(x)$, and $f(x)$, such that (2.106) holds subject to the boundary conditions $u(0) = g$ and $Q(L) = h$. The following features are noteworthy. The weak form has a weakened derivative (i.e., a product of two first derivatives) rather than a second derivative as in the strong form. The symmetry of the derivatives in the first integral in (2.106) will eventually result in a symmetric stiffness matrix for FEA. Finally, a solution to the weak form of the problem is also a solution to the strong form. The weak form, like the strong form, is a *complete* statement of the boundary value problem. A difference is that the weak form is expressed in a way

[2] The prescription $w(0) = 0$ is necessary for this example since an essential boundary condition $u = g$ is imposed at $x = 0$. This restriction will be removed in the more general derivation of §2.3.4 wherein a natural boundary condition is applied at $x = 0$.

that is usually much more amenable than the strong form to approximate methods of solution, for example, numerical methods such as those of FEA.

2.3.3 Interpolation

The next step in application of FEA towards derivation of the element equations requires introduction of scalar functions used to interpolate the solution between nodes within an element. These functions of position x in 1-D problems, denoted by $N_i(x)$, are referred to in various literature as shape functions, interpolation functions, and/or basis functions. We will use the terms shape function and interpolation function interchangeably in this textbook. Summing over n nodes belonging to an element, the displacement field within an element is interpolated as

$$u^{(e)}(x) = \sum_{i=1}^{n} N_i(x) u_i. \tag{2.107}$$

The x-dependence of the primary solution field is thus relegated to the shape functions; nodal displacements u_i are effectively constants. The displacement gradient or derivative, equivalent to the strain in this case, is then obtained directly as

$$\frac{\mathrm{d}u^{(e)}(x)}{\mathrm{d}x} = \frac{\mathrm{d}}{\mathrm{d}x}\left[\sum_{i=1}^{n} N_i(x) u_i\right] = \sum_{i=1}^{n} \frac{\mathrm{d}N_i(x)}{\mathrm{d}x} u_i. \tag{2.108}$$

By definition, shape functions have the following properties: N_i is nonzero only inside elements sharing/containing node i; N_i has a value of one at node i located at point x_i; and at any other node j not located at point x_i, N_i has a value of zero. The latter two properties can be stated succinctly as the identity

$$N_i(x_j) = \delta_{ij}; \qquad \delta_{ij} = 1 \forall i = j, \quad \delta_{ij} = 0 \forall i \neq j. \tag{2.109}$$

The object δ_{ij} is known as the Kronecker delta. The displacement field for the entire domain is determined by the combined contributions of all elements comprising the discretized domain. Since $N_i = 0$ for elements not including or containing node i, the displacement field can be written as

$$u(x) = \sum_{\text{elements}} \left[\sum_{\text{nodes}} N_i(x) u_i\right] = \sum_{i=1}^{p} N_i(x) u_i, \tag{2.110}$$

where now $i = 1, \ldots, p$ is the global node number and p the total number of nodes in the mesh. In this sum, the contribution of each node is counted only once; i.e., shared nodes in neighboring elements are not double-counted.

At this juncture, it is convenient to introduce a notational convention called the summation convention. Summation symbols such as those in (2.107) and (2.108) are cumbersome. The notation can be simplified by invoking a rule that repeated indices are summed. For example,

$$a_i b_i = \sum_{i=1}^{n} a_i b_i = a_1 b_1 + a_2 b_2 + \cdots + a_n b_n. \tag{2.111}$$

Since n is not visible in the expression on the left, the reader must be familiar with the context in which the equation holds. Two contexts encountered in FEA are summation over nodes or degrees-of-freedom, as in (2.107) and (2.108), and summation over spatial dimensions, as in the dot or scalar product of two vectors. Use of the implied summation convention in the latter situation is often referred to as Einstein's summation convention in the realm of tensor analysis [2, 3].

For a bar/truss element with two nodes, the shape functions are linear in x, i.e., displacement is interpolated linearly between the two nodes. It can be verified that the following basic shape functions obey the requisite local properties $N_i(x_j) = \delta_{ij}$ if for an element with nodes at $x = 0, L$ we choose

$$N_1(x) = 1 - \frac{x}{L}, \qquad N_2(x) = \frac{x}{L} \qquad (x \in [0, L]). \tag{2.112}$$

In matrix form, the shape functions are written as the row vector $\boldsymbol{N}$:

$$[\boldsymbol{N}] = [N_1 \quad N_2] = \left[1 - \frac{x}{L} \quad \frac{x}{L}\right]. \tag{2.113}$$

Derivatives of the shape functions comprise the so-called B-matrix $\boldsymbol{B}$:

$$[\boldsymbol{B}] = [B_1 \quad B_2] = \left[\frac{\mathrm{d}N_1}{\mathrm{d}x} \quad \frac{\mathrm{d}N_2}{\mathrm{d}x}\right] = \left[-\frac{1}{L} \quad \frac{1}{L}\right]. \tag{2.114}$$

It follows that, at locations within a linear 1-D mechanical element, displacement (2.107) and strain (2.108) can be expressed, respectively, as

$$u^{(e)} = N_i u_i = [\boldsymbol{N}]\{\boldsymbol{u}\} = [N_1 \quad N_2] \begin{Bmatrix} u_1 \\ u_2 \end{Bmatrix}, \tag{2.115}$$

$$\varepsilon^{(e)} = \frac{\mathrm{d}u^{(e)}}{\mathrm{d}x} = \frac{\mathrm{d}N_i}{\mathrm{d}x} u_i = [\boldsymbol{B}]\{\boldsymbol{u}\} = [B_1 \quad B_2] \begin{Bmatrix} u_1 \\ u_2 \end{Bmatrix}. \tag{2.116}$$

The linear interpolation functions in (2.112) are the simplest available for 1-D elements. Other analytical functions are possible that satisfy the requisite properties of shape functions. For 1-D structures, and many others encountered later in this book, polynomials are the most common class of shape function. Increasing the order of the polynomial corresponds to increasing the number of nodes contained in a single element. For 1-D problems, each additional term in the polynomial correlates with an additional node and one more degree-of-freedom for a given element. Generally, a better approximation to a complex solution field is obtained with higher-order shape functions. Recall from §1.3.1 that increasing the order of element is referred to as p-refinement. However, for the same number of elements in the mesh, i ing the order of interpolation of the shape functions correlates with an increase in the size of the global system of equations, corresponding to greater computational expense.

A general approach towards constructing shape functions for 1-D FEA problems involves so-called Lagrange polynomials. Let n again refer to the total number of nodes or degrees-of-freedom in an element. Let x_i be the coordinate of node i. The Lagrange polynomial is

$$N_i(x) = \prod_{j=1,\ldots,n;j\neq i} \frac{(x-x_j)}{(x_i-x_j)}. \tag{2.117}$$

This formula can be used to generate a set of shape functions of arbitrary order exceeding zero. For example, for an element with three nodes located at (x_1, x_2, x_3), second-order or quadratic shape functions are

$$\begin{aligned} N_1(x) &= \frac{(x-x_2)(x-x_3)}{(x_1-x_2)(x_1-x_3)}, \qquad N_2(x) = \frac{(x-x_1)(x-x_3)}{(x_2-x_1)(x_2-x_3)}, \\ N_3(x) &= \frac{(x-x_1)(x-x_2)}{(x_3-x_1)(x_3-x_2)}. \end{aligned} \tag{2.118}$$

For an element with two nodes at $(x_1, x_2) = (0, L)$, noting $x_2 - x_1 = L$, it is easily verified that the associated Lagrange polynomials are identical to the linear shape functions in (2.112).

2.3.4 *Matrix Equations*

We return to derivation of the governing equations for a single 1-D bar element occupying domain $x \in [0, L]$, with cross-sectional area $A(x)$, elastic modulus $E(x)$, distributed load function $f(x)$, and unknown endpoint displacements $u_1 = u(0), u_2 = u(L)$. Reaction forces at the ends, labeled Q_1 and Q_2, obey

$$Q_1 = -\left(EA\frac{\mathrm{d}u}{\mathrm{d}x}\right)\Big|_{x=0}, \qquad Q_2 = \left(EA\frac{\mathrm{d}u}{\mathrm{d}x}\right)\Big|_{x=L}. \tag{2.119}$$

We follow the same procedure for deriving the strong and weak forms of the equilibrium equations used in §2.3.2, with the exception that now we do not impose $w(0) = 0$, i.e., $w(x)$ is an arbitrary differentiable function. The resulting weak form for static equilibrium requires

$$\int_0^L \left(EA\frac{\mathrm{d}w}{\mathrm{d}x}\frac{\mathrm{d}u}{\mathrm{d}x}\right)\mathrm{d}x = Q_1 w(0) + Q_2 w(L) + \int_0^L wf\,\mathrm{d}x. \tag{2.120}$$

We now make the following substitutions using general shape functions $N_i(x)$ [or $N_j(x)$] and invoking summation over nodes i [or j] in the element:

$$u(x) = N_j(x)u_j, \qquad w(x) = N_i(x)w_i, \tag{2.121}$$

where w_i are constant values of $w(x)$ at nodal points x_i. The weak form expression thus becomes

$$\int_0^L \left(EA\frac{\mathrm{d}N_i}{\mathrm{d}x}\frac{\mathrm{d}N_j}{\mathrm{d}x} \right) u_j w_i \mathrm{d}x = Q_1 N_i(0) w_i + Q_2 N_i(L) w_i + \int_0^L f N_i w_i \mathrm{d}x. \tag{2.122}$$

Since $w(x)$ is arbitrary, the weak form must be valid for any selected combination of constants w_i. Therefore, w_i can be factored out of the weak form expression (2.122) such that the remaining system of n equilibrium equations becomes

$$\int_0^L \left(EA\frac{\mathrm{d}N_i}{\mathrm{d}x}\frac{\mathrm{d}N_j}{\mathrm{d}x} \right) u_j \mathrm{d}x = Q_1 N_i(0) + Q_2 N_i(L) + \int_0^L f N_i \mathrm{d}x \qquad (i = 1, \ldots, n), \tag{2.123}$$

with n the number of nodes in the element. In other words, the number of simultaneous equations to be solved equals the number of possible values of index i.

Straightforward substitution of definitions of shape function matrix $\boldsymbol{N}$, shape function derivative matrix $\boldsymbol{B}$, and displacement vector $\boldsymbol{u}$ then enable transformation of (2.123) into vector-matrix form:

$$\int_0^L EA[\boldsymbol{B}]^\mathrm{T}[\boldsymbol{B}]\{\boldsymbol{u}\}\mathrm{d}x = [\boldsymbol{N}]^\mathrm{T}_{|x=0} Q_1 + [\boldsymbol{N}]^\mathrm{T}_{|x=L} Q_2 + \int_0^L f[\boldsymbol{N}]^\mathrm{T}\mathrm{d}x. \tag{2.124}$$

Since $\boldsymbol{u}$ is a vector of effectively constant nodal displacements, it can be moved out of the integral on the left side of the equality without consequence, and we may write

$$\int_0^L EA[\boldsymbol{B}]^\mathrm{T}[\boldsymbol{B}]\{\boldsymbol{u}\}\mathrm{d}x = \left(\int_0^L EA[\boldsymbol{B}]^\mathrm{T}[\boldsymbol{B}]\mathrm{d}x \right) \{\boldsymbol{u}\} = [\boldsymbol{k}^{(e)}]\{\boldsymbol{u}^{(e)}\}. \tag{2.125}$$

The general integral expression for the element stiffness matrix is thus

$$[\boldsymbol{k}^{(e)}] = \int_{\Omega^{(e)}} EA[\boldsymbol{B}]^\mathrm{T}[\boldsymbol{B}]\mathrm{d}x. \tag{2.126}$$

We now make the following additional simplifying assumptions for the element: the element is linear ($i = 1, 2$) with shape functions (2.112); the properties A and E are constant over the element; and f is a constant. From (2.114), shape function derivatives are now constants, so the integral equations in (2.124) reduce to

$$EA[\boldsymbol{B}]^\mathrm{T}[\boldsymbol{B}]\{\boldsymbol{u}\} \int_0^L \mathrm{d}x = [\boldsymbol{N}]^\mathrm{T}_{|x=0} Q_1 + [\boldsymbol{N}]^\mathrm{T}_{|x=L} Q_2 + f \int_0^L [\boldsymbol{N}]^\mathrm{T}\mathrm{d}x. \tag{2.127}$$

Performing the integration and matrix multiplication operations then yields

$$\frac{EA}{L}\begin{bmatrix}1 & -1\\ -1 & 1\end{bmatrix}\begin{Bmatrix}u_1\\ u_2\end{Bmatrix}=\begin{Bmatrix}Q_1\\ Q_2\end{Bmatrix}+\frac{fL}{2}\begin{Bmatrix}1\\ 1\end{Bmatrix}. \tag{2.128}$$

The 2×2 matrix on the left side is the element stiffness $\boldsymbol{k}^{(e)}$. Noting that $k = EA/L$, the set of equations in (2.128) is identical to that derived using the Direct Method (DM) in (2.13) when $f = 0$; however, the distributed load vector could have been incorporated into the DM, which we recall follows an intuitive process that becomes cumbersome for more complex physics. In contrast, given the differential equation(s) in weak form, the derivation of the element-level matrix equations using FEA follows the same systematic process irrespective of physics: determine the shape functions and their derivatives, evaluate the force terms due to nodal and distributed loads, and evaluate the element stiffness matrix. The latter two evaluations require integration of shape functions and shape function derivatives over the element. It is remarked that although the element equations derived for linear interpolation functions and constant bar properties E and A are identical when obtained from either DM or FEM, *they are not necessarily identical in the general case when different interpolation functions and/or spatially variable properties apply.*

With the element-level equilibrium equations now at hand, assembly of the global system of equations follows the identical procedure given in §2.2.2 in the context of DM. Similarly, reduction of the global system using boundary conditions, followed by solution of the global system for primary variables (u_i in this case), invokes the identical processes discussed in §2.2. Problems involving 2-D space trusses, 3-D space trusses, and thermal strains can be approached in a straightforward manner using techniques discussed already in §2.2.8, §2.2.9, and §2.2.10, respectively. The only remaining noteworthy difference in analysis of a 1-D mechanical boundary value problem via FEM versus DM pertains to post-processing.

2.3.5 Post-processing

Assume that the solution of the global system results in the nodal displacement vector $\boldsymbol{u}$, which as usual for static problems is obtained from

$$\{\boldsymbol{u}\} = [\boldsymbol{K}]^{-1}\{\boldsymbol{F}\}. \tag{2.129}$$

Vector $\boldsymbol{F}$ contains combined contributions of distributed loads and natural boundary conditions at the nodes. For 1-D problems involving mechanical bar or truss elements, variables usually sought during post-processing are internal forces supported in each element, strains within each element, and stresses within each element. Internal forces are obtained via use of the element equilibrium equations derived in §2.3.4:

$$\{\boldsymbol{Q}^{(e)}\} = [\boldsymbol{k}^{(e)}]\{\boldsymbol{u}^{(e)}\}. \tag{2.130}$$

Superscripts in parentheses again refer to element numbers. Referring to (2.108), strain at any location x in an element can be calculated from displacement via use of the shape function derivative matrix:

$$\varepsilon^{(e)}(x) = \frac{\mathrm{d}u^{(e)}}{\mathrm{d}x}(x) = [\boldsymbol{B}^{(e)}(x)]\{\boldsymbol{u}^{(e)}\}, \tag{2.131}$$

where the B-matrix is evaluated at point x. This expression differs from that in DM; the Direct Method does not invoke shape functions or their derivatives. Stress supported by the bar or truss at point x then follows directly from the linear elastic constitutive model:

$$\sigma^{(e)}(x) = E(x)\varepsilon^{(e)}(x) = E(x)[\boldsymbol{B}^{(e)}(x)]\{\boldsymbol{u}^{(e)}\}. \tag{2.132}$$

2.3.6 Examples

Two example applications of 1-D FEM are discussed in turn. The first problem is defined as follows. A bar of total length $3L$ consists of three elements in series, each of identical length L, constant modulus E, and uniform cross-sectional area A. The coordinate system is defined such that $x \in [0, 3L]$, with the essential boundary condition at node 1 given by $u_1 = u(x_1) = u(0) = 0$. Natural boundary conditions on the remaining three nodes at $(x_2, x_3, x_4 = L, 2L, 3L)$ are $Q_2 = -5EA$, $Q_3 = 0$, and $Q_4 = 10EA$ in consistent units. We seek the unknown nodal displacements at (x_2, x_3, x_4) and the stress in element 1 which covers the region $x \in [0, L]$.

Analysis begins with the element equation of (2.127). Assembly of this equation for three elements in series and application of the boundary conditions results in

$$\frac{EA}{L}\begin{bmatrix} 1 & -1 & 0 & 0 \\ -1 & 2 & -1 & 0 \\ 0 & -1 & 2 & -1 \\ 0 & 0 & -1 & 1 \end{bmatrix}\begin{Bmatrix} 0 \\ u_2 \\ u_3 \\ u_4 \end{Bmatrix} = \begin{Bmatrix} Q_1 \\ -5EA \\ 0 \\ 10EA \end{Bmatrix}. \tag{2.133}$$

We notice that the global stiffness is symmetric, tridiagonal, and of size 4×4 since the problem contains 4 degrees of freedom. Notice also that Q_1 is unknown because $u_1 = 0$ is imposed. Invoking the latter (essential) boundary condition, the first row and column of the global stiffness can be safely eliminated without consequence, leading to a condensed system of 3 equations and 3 unknown nodal displacements:

$$\frac{EA}{L}\begin{bmatrix} 2 & -1 & 0 \\ -1 & 2 & -1 \\ 0 & -1 & 1 \end{bmatrix}\begin{Bmatrix} u_2 \\ u_3 \\ u_4 \end{Bmatrix} = \begin{Bmatrix} -5EA \\ 0 \\ 10EA \end{Bmatrix}. \tag{2.134}$$

Inverting the condensed stiffness matrix and solving the resulting system of three equations gives

$$\begin{Bmatrix} u_2 \\ u_3 \\ u_4 \end{Bmatrix} = L\begin{Bmatrix} 5 \\ 15 \\ 25 \end{Bmatrix}. \tag{2.135}$$

Finally, we obtain the stress in element 1, which is constant over the element (why?), via use of (2.132):

$$\sigma^{(1)} = E[\boldsymbol{B}^{(1)}]\{\boldsymbol{u}^{(1)}\} = E[-1/L \quad 1/L]\begin{Bmatrix} 0 \\ 5L \end{Bmatrix} = 5E. \tag{2.136}$$

We now consider a second 1-D problem, here consisting of a single bar element occupying the domain $x \in [0, L]$. The element has a constant modulus E but a variable cross-section with area $A(x) = A_0(1 + x/L)$, where A_0 is a positive constant. A concentrated force applied at $x = L$ has magnitude P, so nodal reaction forces are $Q_2 = -Q_1 = P$. We seek the stiffness matrix $\boldsymbol{k}^{(e)}$ for this non-uniform element obtained using the FEM and wish to compare the result with the exact stiffness obtained analytically.

The approach from FEA requires use of (2.126). We choose linear interpolation functions in (2.112) with constant derivatives (2.114). The element stiffness matrix in the FEM approximation becomes

$$\begin{aligned}
[\boldsymbol{k}^{(e)}] &= \int_0^L EA[\boldsymbol{B}]^{\mathrm{T}}[\boldsymbol{B}]\mathrm{d}x = E[\boldsymbol{B}]^{\mathrm{T}}[\boldsymbol{B}]\int_0^L A\mathrm{d}x \\
&= \frac{EA_0}{L^2}\begin{bmatrix} 1 & -1 \\ -1 & 1 \end{bmatrix}\int_0^L \left(1 + \frac{x}{L}\right)\mathrm{d}x = \frac{3EA_0}{2L}\begin{bmatrix} 1 & -1 \\ -1 & 1 \end{bmatrix} = k_{\mathrm{FEM}}\begin{bmatrix} 1 & -1 \\ -1 & 1 \end{bmatrix},
\end{aligned} \tag{2.137}$$

where $k_{\mathrm{FEM}} = 1.5EA_0/L$. To obtain the analytical solution, we note that the stress at point x is

$$\sigma(x) = \frac{P}{A(x)} = E\frac{\mathrm{d}u(x)}{\mathrm{d}x}. \tag{2.138}$$

Therefore,

$$\mathrm{d}u = \frac{P\mathrm{d}x}{EA(x)} \Rightarrow u_2 - u_1 = \int_0^L \frac{P\mathrm{d}x}{EA(x)} = \int_0^L \frac{P\mathrm{d}x}{EA_0(1 + x/L)} = \frac{PL}{EA_0}\ln 2. \tag{2.139}$$

Noting that $Q_2 = -Q_1 = P$, we can write (2.139) in vector-matrix form as

$$\frac{1}{\ln 2}\frac{EA_0}{L}\begin{bmatrix} 1 & -1 \\ -1 & 1 \end{bmatrix}\begin{Bmatrix} u_1 \\ u_2 \end{Bmatrix} = \begin{Bmatrix} -Q \\ Q \end{Bmatrix}. \tag{2.140}$$

Therefore, the exact stiffness matrix is

$$[\boldsymbol{k}^{(e)}] = \frac{1}{\ln 2}\frac{EA_0}{L}\begin{bmatrix} 1 & -1 \\ -1 & 1 \end{bmatrix} = 1.443\frac{EA_0}{L}\begin{bmatrix} 1 & -1 \\ -1 & 1 \end{bmatrix} = k_{\mathrm{exact}}\begin{bmatrix} 1 & -1 \\ -1 & 1 \end{bmatrix}, \tag{2.141}$$

with $k_{\mathrm{exact}} = 1.443EA_0/L \approx 0.96k_{\mathrm{FEM}}$. These results imply numerical error in each component of the FEM stiffness in (2.137) of around 4%. With P imposed, this would translate to an error of the same order of magnitude in nodal displacement u_2 since $u_2 \propto 1/k$.

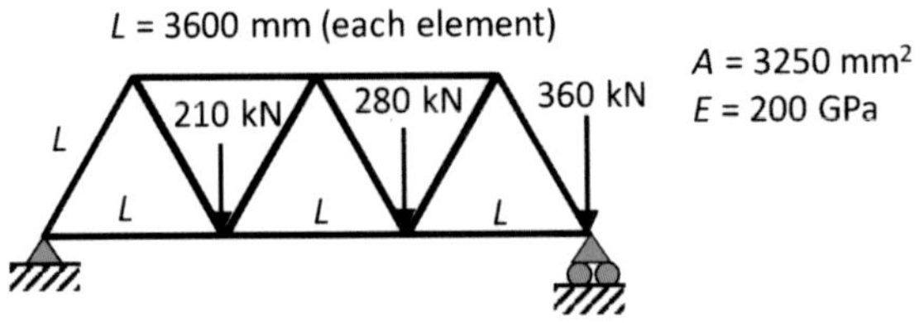

Fig. 2.11 Truss bridge example problem: geometry, boundary conditions, and properties

2.4 ANSYS Example: Truss Bridge

Presented here is the first example in this book of analysis of a problem using FE software. The package used for this, and all, such examples in this book is ANSYS Mechanical Release 17.2, SAS IP, Inc., 2016. Another comprehensive reference on use of the ANSYS Graphical User Interface (GUI), albeit an earlier software release (2006), with step-by-step command instructions for a number of examples is [4]. Inevitably, some software features are updated and changed as new versions are released.

2.4.1 Problem Statement

The physical problem is illustrated in Fig. 2.11 and is adapted from ANSYS tutorials posted by the University of Alberta. A bridge consisting of eleven truss members is joined at seven points. Each member is of equal length $L = 3.6$ m, cross-sectional area $A = 3250\ \text{mm}^2$, and Young's modulus $E = 200$ GPa. The bridge is completely pinned at the lower left corner to prevent rigid body motion. At the lower right corner, a roller prevents vertical displacement ($u_y = 0$) but permits horizontal motion, i.e., $u_x \neq 0$ in general. Three forces are applied of magnitudes 210 kN, 280 kN, and 360 kN. The direction of each force is strictly vertical and downward, i.e., in the $-y$-direction.

The objective is determination of the primary solution: the displacement components u_x and u_y for each joint that will be represented by a node. A secondary objective is qualitative validation of the solution via visualization of the deformed shape using the software graphics tools.

2.4.2 Pre-processing

Pre-processing here includes selection of element type and material properties, followed by mesh creation. The analysis in the ANSYS package is initiated by selec-

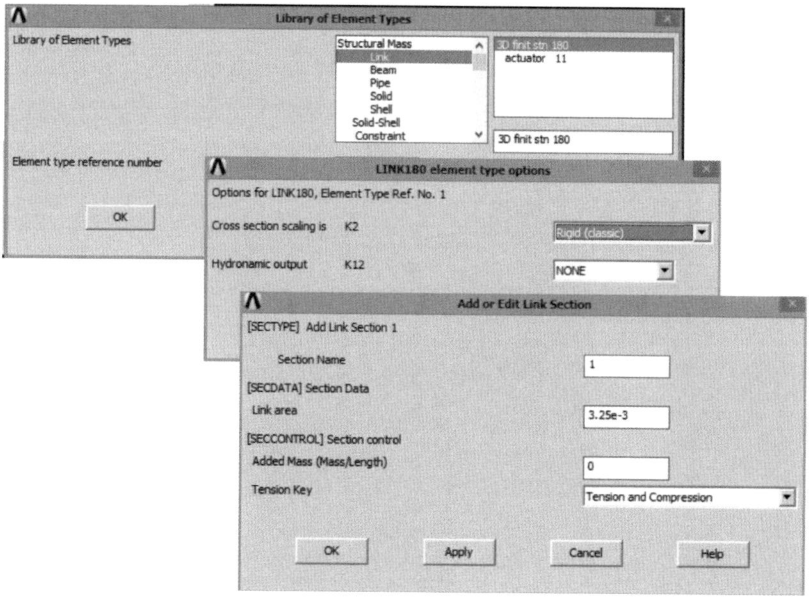

Fig. 2.12 Truss bridge example problem: element data

tion of the element type suitable for a linear truss, the `link 180` element, using the sequence below:

```
Main Menu > Preprocessor > Element Type > Add/Edit/Delete
> Add > Link > 3D finit stn 180
```

This element enables consideration of nonlinearities associated with large deformation, but it remains suitable for linear problems. A section must be created in this software package:

```
Main Menu > Preprocessor > Sections > Link > Add
```

Enter the area $A = 3.25\text{e}{-3}\ \text{m}^2$, as shown in Fig. 2.12, and ensure that the elements enable both tension and compression stiffness[3]. The name of the section can be entered arbitrarily; since all elements have the same properties, all will be assigned to this section by default. Next, the material model is selected, specifically isotropic linear elasticity:

[3] A truss element with only tensile stiffness would be physically representative of a rope or string.

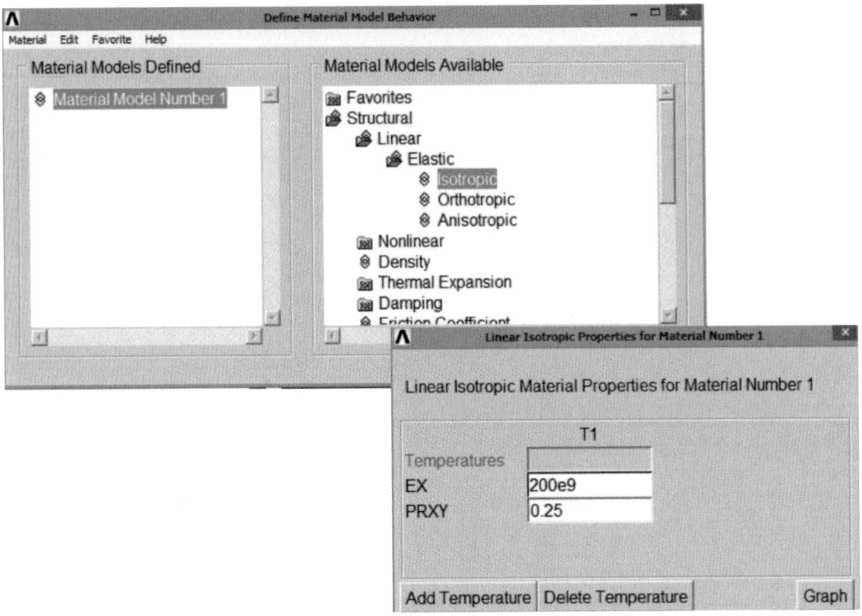

Fig. 2.13 Truss bridge example problem: material data

```
Main Menu > Preprocessor > Material Properties > Material
Models > Structural > Linear > Elastic > Isotropic
```

The software then prompts for entry of the elastic modulus E and Poisson's ratio ν. The former is prescribed as 200e9 Pa, while the latter is arbitrary for this problem and can be safely chosen as 0.25, for example, though any value obeying $-1 < \nu < 0.5$ would be sufficient. Notice that we use SI units for consistency. The procedure is shown in Fig. 2.13. The next step is creation of the mesh. For this problem, there is no need to create a separate set of geometric entities prior to mesh creation. Rather, the most efficient method is creation of the nodes and elements directly. Nodes are produced by the following sequence:

```
Main Menu > Preprocessor > Modeling > Create > Nodes > In
Active CS
```

Enter a node number and coordinates for each point corresponding to intersection of the truss members in Fig. 2.11. The procedure and results after all nodes have been created are shown in Fig. 2.14. Then, elements are created directly via

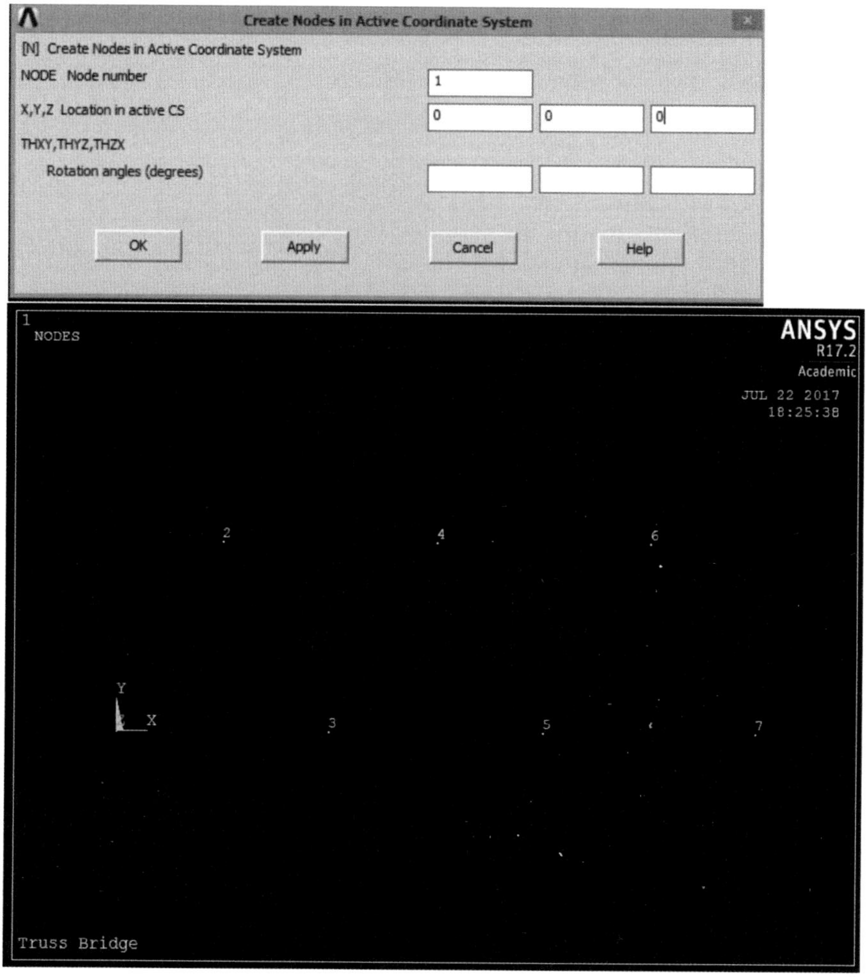

Fig. 2.14 Truss bridge example problem: node creation

```
Main Menu > Preprocessor > Modeling > Create > Elements >
User Numbered > Thru Nodes
```

Numbered elements are depicted in Fig. 2.15.

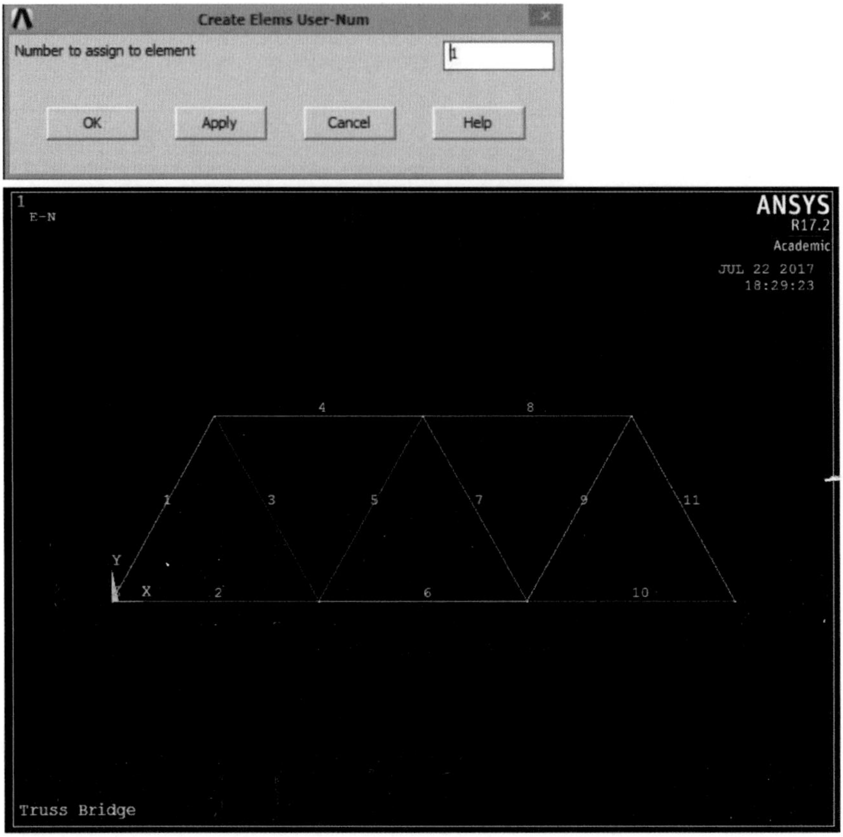

Fig. 2.15 Truss bridge example problem: element creation

2.4.3 Solution

The solution steps for this problem define the problem type, boundary conditions, and possible solver options. The analysis type is static, invoked by the following sequence:

```
Main Menu > Solution > Analysis Type > New Analysis >
Static
```

See Fig. 2.17. Boundary conditions are enabled here in the solution module, though they could equivalently be prescribed in the pre-processing module. First, displacement conditions, i.e., essential BCs, are enforced at nodes 1 and 7:

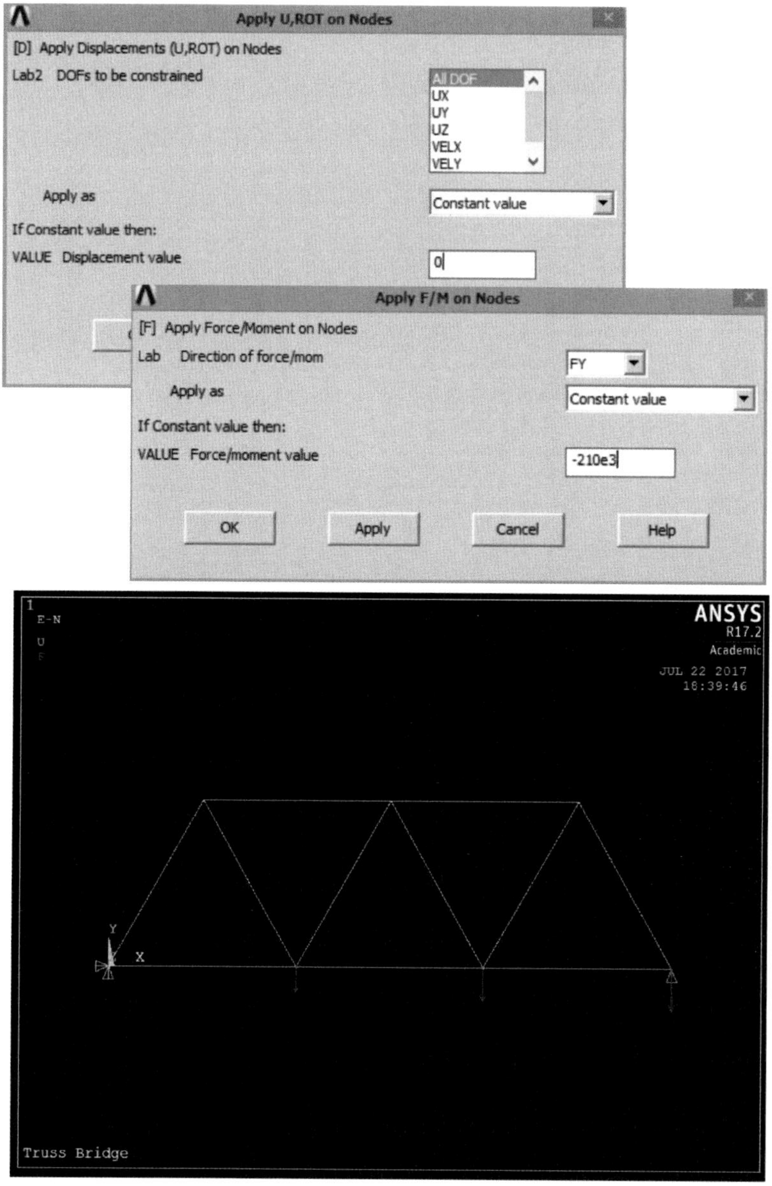

Fig. 2.16 Truss bridge example problem: boundary conditions

```
Main Menu > Solution > Define Loads > Apply > Structural >
Displacement > On Nodes
```

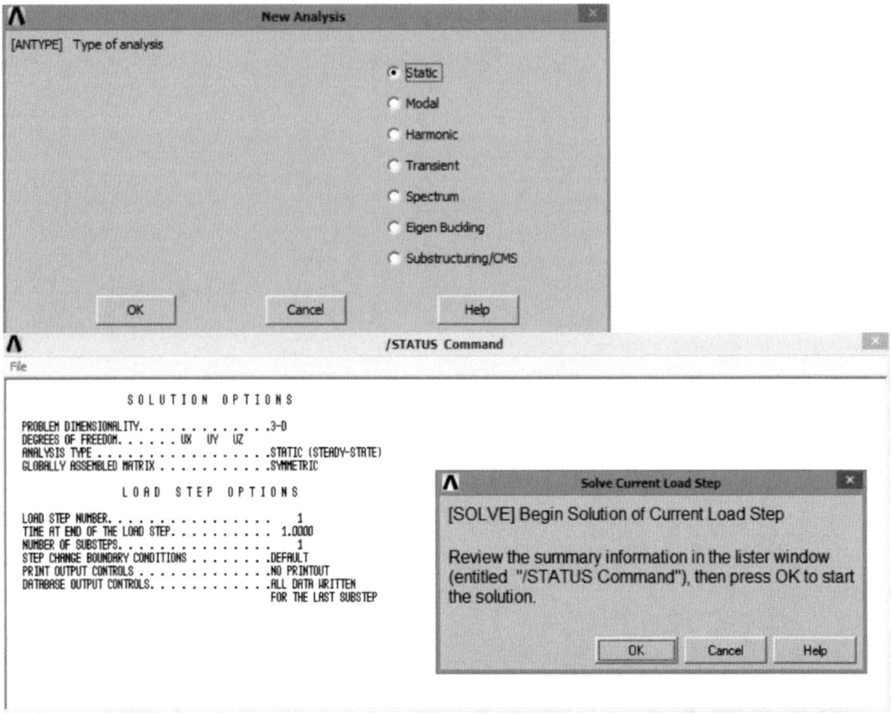

Fig. 2.17 Truss bridge example problem: analysis type and solver pre-check

Then forces, i.e., natural boundary conditions corresponding to loads in Fig. 2.11, are applied to nodes 3, 5, and 7:

```
Main Menu > Solution > Define Loads > Apply > Structural >
Force/Moment > On Nodes
```

Boundary conditions are visualized in Fig. 2.16. One of the two conditions on the rightmost node (node 7) is redundant as will be discussed in §2.4.4. The default solver options are enabled for this problem. These are displayed by the pre-check that is enacted by the sequence below:

```
Main Menu > Solution > Solve > Current LS > OK
```

Click OK to execute the solver and obtain the solution as indicated in Fig. 2.17.

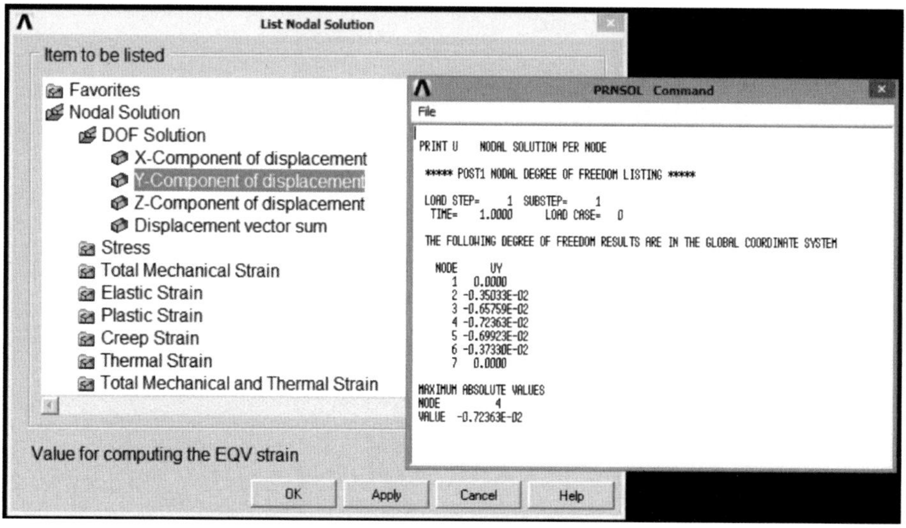

Fig. 2.18 Truss bridge example problem: primary solution

2.4.4 Post-processing

The primary solution is the displacement vector $\boldsymbol{u}$ with components u_x and u_y for each of the seven nodes. The following GUI selections create a list of the vertical displacement components, u_y, for each node:

```
Main Menu > General Postproc > List Results > Nodal
Solution > DOF Solution > Y-Component of displacement
```

See Fig. 2.18 for the associated results. Horizontal components can be viewed in a similar fashion. The deformed shape of the truss, i.e., the structure with nodes displaced by $\boldsymbol{u}$, can be viewed in superposition with the original unloaded structure as follows:

```
Main Menu > General Postproc > Plot Results > Deformed
Shape
```

Notice that the displacements are magnified by default in the ANSYS GUI to enable visualization. Numerical values are small, with a maximum vertical displacement on the order of 0.2% of L, which would be invisible in the graphics window if not magnified. Qualitatively, the results agree with physical intuition. The largest downward displacement occurs at node 4, at the middle and top of the bridge (Fig. 2.14). At node 7, horizontal displacement occurs but vertical displacement does not.

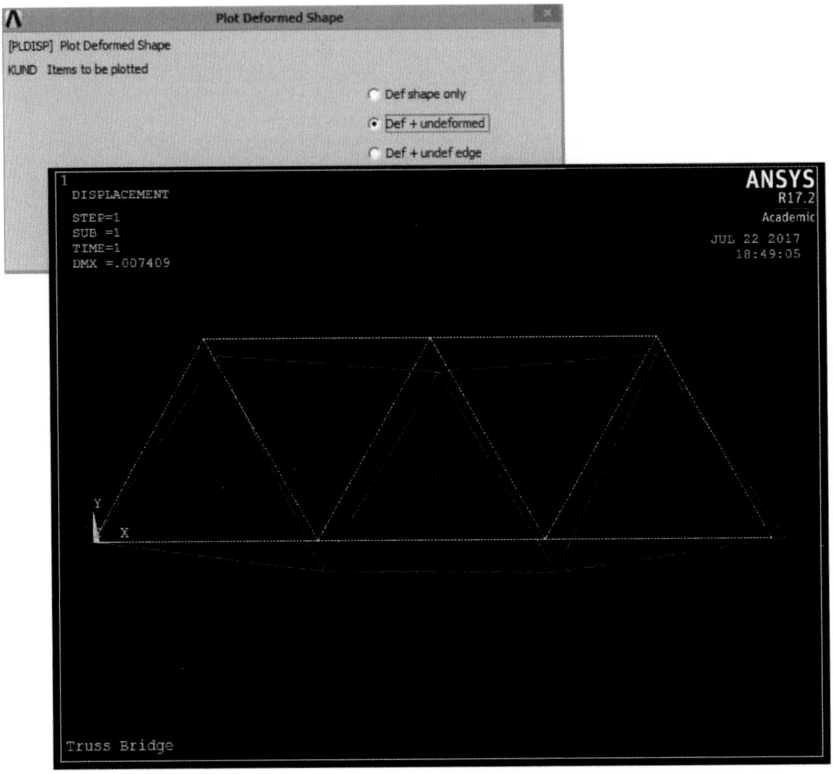

Fig. 2.19 Truss bridge example problem: deformed shape

The force of magnitude 360 kN acting on this rightmost node has no effect on the solution. As remarked earlier in §2.2.5, prescription of an essential and a natural boundary condition that are work conjugate to each other at the same node is physically inappropriate. Here, since $u_y = 0$ at node 7, ANSYS simply ignores the redundant conjugate point force of 360 kN. To prove this, remove the force from the model and repeat the solution steps. The resulting nodal displacements will be identical to those computed already with the nonzero force in effect.

2.5 Summary

This chapter has derived the equilibrium equations for 1-D mechanical problems via two approaches: DM and FEM. The former approach is informal, requiring substantial intuition from the analyst regarding physics of the problem. The latter approach is more formal, relying on standardized mathematical procedures given the strong form of the governing differential equation(s). It has been shown how, for certain simple situations, the element equations derived via DM and FEM coincide, though this is not true for all problems. Procedures common among the two methods include assembly of the global system from the element equations, application of boundary conditions, and solution of the condensed set of global equations.

This chapter has also considered post-processing in DM and FEM, analysis of 2-D and 3-D space trusses composed of 1-D elements, as well as some basic concepts from thermoelasticity. In later chapters, the element equations will always be derived via FEM, foregoing the direct approach which tends to be unwieldy for more complicated physical problems.

2.6 Problems

2.6.1. A single 3-node quadratic truss element of length $2L$ has nodes located at $x_1 = -L$, $x_2 = 0$, and $x_3 = L$. In the following, assume Young's modulus E and cross-sectional area A are known constants. The shape functions for this element are

$$N_1 = \frac{1}{2L^2}x(x-L), \qquad N_2 = \frac{1}{L^2}(L^2-x^2), \qquad N_3 = \frac{1}{2L^2}x(x+L).$$

a. Show that the shape functions have the properties that at a node j, $N_i = 1$ when $i = j$ and $N_i = 0$ when $i \neq j$.
b. Determine the shape function derivatives that enter the $[\boldsymbol{B}]$ matrix.
c. Determine the element stiffness matrix by integrating the product $[\boldsymbol{B}]^{\mathrm{T}}EA[\boldsymbol{B}]$ over the length of the truss.
d. Compare the stiffness matrix from part **c.** with the global stiffness matrix for two assembled linear truss elements, each of length L. Describe how and why they may or may not differ.

2.6.2. Solve the 5-spring problem as described in §2.2.3, but now use 5 linear truss elements in place of springs. Each element is of the same length, but elements 1 and 5, with $E = 125$ GPa and $A = 0.04$ m^2, have different properties than elements 2, 3, and 4 with $E = 250$ GPa and $A = 0.01$ m^2. Assume all displacements are directed only parallel along the $\pm x$-axis. If each truss element is of length $L = 0.2$ m and the force applied is $P = 30$ kN, determine the following:
a. The global stiffness matrix and force-displacement equations in symbolic form (i.e., in terms of E's, A's, L's, Q's, u's) by assembling the element equations; you do not have to re-derive the element equations already described in §2.2 and §2.3.
b. The condensed system of equations after applying boundary conditions.
c. The displacements at the nodes.
d. The (internal) forces from the elements at the nodes.
e. The stresses and strains in each element found using $[\boldsymbol{B}]$.
f. Solve the problem using any commercial FEM software and verify your numerical results via comparison with your solutions to parts **c.** and **d.** Obtain screen-captured

images or exported graphics files of the GUI window(s) showing displacements, forces, stresses, and strains.

References

1. G. Strang, *Introduction to Linear Algebra* (Wellesley-Cambridge Press, Wellesley, MA, 1993)
2. J.D. Clayton, *Nonlinear Mechanics of Crystals* (Springer, Dordrecht, 2011)
3. J.D. Clayton, *Differential Geometry and Kinematics of Continua* (World Scientific, Singapore, 2014)
4. E. Madenci and I. Guven, *The Finite Element Method and Applications in Engineering Using ANSYS* (Springer, New York, 2006)

Chapter 3
Beams and Frames

Abstract The finite element method is developed for mechanical elements that support shearing and bending loads, specifically beam elements. Euler-Bernoulli beam theory is summarized, including a derivation of the governing fourth-order ordinary differential equation for static equilibrium. The weak form of this equation is derived. Shape functions are introduced for beam elements, and then element-level equations of equilibrium are developed in vector-matrix form. Assembly of the global equations, boundary conditions, solution methods, and post-processing are discussed. Several example beam problems are analyzed. Frame elements, which can be considered a superposition of two-dimensional truss/bar elements with Euler-Bernoulli beam elements, are described.

Chapter 2 first develops the finite element method (FEM) for geometrically 1-D elements that can undergo transverse deflection and bending, i.e., beam elements. This is followed by development of FEM for frame elements (here in 2-D), which can undergo axial deformation in addition to transverse deflection and bending. The underlying theory that will lead to the local equilibrium equations for Euler-Bernoulli beams is described next.

3.1 Euler-Bernoulli Beam Theory

This textbook develops FEM for the simplest beam theory of engineering mechanics: Euler-Bernoulli beam theory. The governing equations are those presented in standard undergraduate engineering textbooks on strength of materials [1]. This theory is deemed most suitable for slender beams undergoing relatively small deflections since shear strain is omitted and since a linear elastic material response is assumed.

The kinematics of Euler-Bernoulli beam theory are illustrated in Fig. 3.1. The reference configuration refers to the beam prior to deflection or bending. The geometry of the beam is said to be 1-D since the reference configuration is fully described

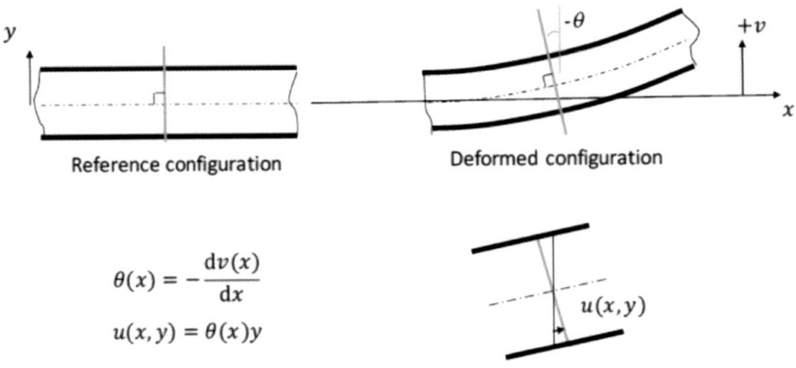

Fig. 3.1 Euler-Bernoulli beam kinematics

by a single coordinate (x); however, deformations are 2-D since they exist in the xy-plane. The neutral axis is the x-axis, and the shear deflection υ is positive when directed in the $+y$-direction. The rotation or slope of the beam is denoted by θ, positive in sign when clockwise in sense, such that for an upwardly sloping beam as shown in Fig. 3.1, the sign of θ is negative. Axial deformation u results purely from rotation of cross-sections about the neutral axis; along the neutral axis, the beam undergoes no stretching or compression. In other words, the key assumptions for kinematics of Euler-Bernoulli beams are

- Plane cross-sections remain planar;
- Cross-sections stay normal to the neutral plane;
- Axial deformation is the product slope × distance from the neutral axis.

Mathematically, the governing kinematic relations of Euler-Bernoulli beam theory can be stated as follows:

$$\upsilon = \upsilon(x), \tag{3.1}$$

$$\theta(x) = -\frac{\mathrm{d}\upsilon(x)}{\mathrm{d}x}, \tag{3.2}$$

$$u(x,y) = \theta(x)y = -y\frac{\mathrm{d}\upsilon(x)}{\mathrm{d}x}. \tag{3.3}$$

Also introduced are the curvature κ and axial strain ε, the latter positive (negative) in sign for tension (compression):

$$\kappa(x) = \frac{\mathrm{d}\theta(x)}{\mathrm{d}x} = -\frac{\mathrm{d}^2\upsilon(x)}{\mathrm{d}x^2}, \tag{3.4}$$

$$\varepsilon(x,y) = \frac{\partial u(x,y)}{\partial x} = y\frac{\mathrm{d}\theta(x)}{\mathrm{d}x} = -y\frac{\mathrm{d}^2 \upsilon(x)}{\mathrm{d}x^2} = y\kappa(x). \tag{3.5}$$

All other components of the strain tensor vanish identically. The above relations—in particular (3.2) and (3.5)—can be interpreted as compatibility conditions for Euler-Bernoulli beams.

The constitutive model is that of linear isotropic elasticity. Letting σ denote the axial stress, positive in tension, and letting E denote the elastic modulus which here may depend on x but not y,

$$\sigma(x,y) = E(x)\varepsilon(x,y) = E(x)y\kappa(x). \tag{3.6}$$

The bending moment M at an axial point x on the beam is the negative integral of the stress couple over the cross-sectional area A of the beam:

$$\begin{aligned} M(x) &= -\int_A \sigma(x,y)y\mathrm{d}A = -\int_A E(x)y\kappa(x)\mathrm{d}y\mathrm{d}z = \frac{\mathrm{d}^2 \upsilon(x)}{\mathrm{d}x^2}E(x)\int_A y^2\mathrm{d}y\mathrm{d}z \\ &= E(x)I(x)\frac{\mathrm{d}^2 \upsilon(x)}{\mathrm{d}x^2}, \end{aligned} \tag{3.7}$$

where the moment of inertia I is

$$I(x) = \int_A y^2\mathrm{d}y\mathrm{d}z = \int_A y^2\mathrm{d}A. \tag{3.8}$$

The shearing force V is defined as the derivative of the bending moment:

$$V(x) = \frac{\mathrm{d}M(x)}{\mathrm{d}x} = \frac{\mathrm{d}}{\mathrm{d}x}\left[E(x)I(x)\frac{\mathrm{d}^2 \upsilon(x)}{\mathrm{d}x^2}\right]. \tag{3.9}$$

Possible boundary conditions that may be applied at a point x consist of the following:

- **Essential boundary conditions**: deflection $\upsilon(x)$ and/or rotation $\theta(x)$;
- **Natural boundary conditions**: shear force $V(x)$ and/or moment $M(x)$.

Work conjugate pairs are (υ, V) and (θ, M). Only one entry of each pair may be imposed at any single point x. For example, one cannot impose both M and θ simultaneously at the same point x. Some permissible example combinations of beam boundary conditions are shown in Fig. 3.2. Also admissible in Euler-Bernoulli beam theory is a distributed load (force per unit length) denoted by $f(x)$, acting in the $\pm y$ direction, with the sign of f defined herein as positive for loading in the upward $(+y)$ sense.

Physical Description	Essential Condition	Natural Condition
Free	None	All
Pinned	$v = 0$	Moment
Roller (w/ bar DOF)	$v = 0$	Moment
Clamped	$v = 0$ $\theta = 0$	None

Fig. 3.2 Some common boundary conditions for Euler-Bernoulli beams

3.2 Strong and Weak Forms

The governing Ordinary Differential Equation (ODE) entering the strong form of the problem statement for an Euler-Bernoulli beam is derived next. Consider the beam shown in Fig. 3.3, which is subjected to a distributed load $f(x)$, a concentrated shearing force V_0 at $x = x_V$, a concentrated bending moment M_0 at $x = x_M$, a rotation at $x = x_\theta$, and a transverse displacement at $x = x_v$. Two equilibrium equations, i.e., two conservation laws, will be derived for the differential element on the right side of Fig. 3.3 and then combined to produce a single strong form differential equation. The first equilibrium equation is the balance of linear momentum in the y-direction, which corresponds in the present quasi-static case to a balance of transverse forces:

$$V(x) - V(x + \mathrm{d}x) + f(x + \mathrm{d}x)\mathrm{d}x = 0. \tag{3.10}$$

Dividing this equation by $\mathrm{d}x$ and taking the limit $\mathrm{d}x \to 0$ produces

$$\frac{\mathrm{d}V(x)}{\mathrm{d}x} = f(x). \tag{3.11}$$

The second equilibrium equation is the balance of angular momentum about the z-axis, where bending moments are here taken as positive in sign when counterclockwise in sense:

$$M(x + \mathrm{d}x) - M(x) - V(x)\mathrm{d}x - f(x + \mathrm{d}x)\mathrm{d}x\frac{\mathrm{d}x}{2} = 0 \tag{3.12}$$

Again dividing by $\mathrm{d}x$ and letting $\mathrm{d}x \to 0$, it follows that

$$\frac{\mathrm{d}M(x)}{\mathrm{d}x} = V(x). \tag{3.13}$$

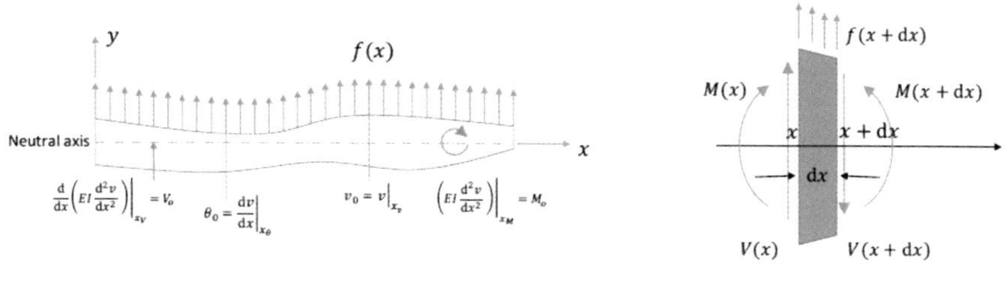

Fig. 3.3 Euler-Bernoulli beam (left) and its differential element (right)

Substituting (3.13) into (3.11) gives

$$\frac{\mathrm{d}^2M(x)}{\mathrm{d}x^2} = f(x). \tag{3.14}$$

Finally, using (3.7), the strong form ODE for Euler-Bernoulli beam theory is, in terms of primary field variable deflection $\upsilon(x)$,

$$\frac{\mathrm{d}^2}{\mathrm{d}x^2}\left[E(x)I(x)\frac{\mathrm{d}^2\upsilon(x)}{\mathrm{d}x^2}\right] = f(x). \tag{3.15}$$

The strong form of the boundary value problem can be stated as follows. Find the deflection function $\upsilon(x)$ and its derivative $\theta(x) = -\mathrm{d}\upsilon(x)/\mathrm{d}x$ that satisfy (3.15), given beam properties $E(x)$, $I(x)$, and L, where $x \in [0,L]$, and subject to the boundary conditions at $x = 0, L$ and a distributed load function $f(x)$, a concentrated shearing force V_0 at $x = x_V$, a concentrated bending moment M_0 at $x = x_M$, an imposed rotation at $x = x_\theta$, and an imposed transverse displacement at $x = x_\upsilon$. Notice that (3.15) can be considered a fourth-order ODE in υ.

To derive the weak form of (3.15), a weight function $w(x)$, twice differentiable but otherwise arbitrary, is introduced on the domain $\{\Omega : x \in [0,L]\}$. Multiplying (3.15) by w and then integrating over Ω gives

$$\int_0^L \left(w(x)\frac{\mathrm{d}^2}{\mathrm{d}x^2}\left[E(x)I(x)\frac{\mathrm{d}^2\upsilon(x)}{\mathrm{d}x^2}\right] - w(x)f(x) \right) \mathrm{d}x = 0. \tag{3.16}$$

Integrating by parts once then produces

$$-\int_0^L \left(\frac{\mathrm{d}w}{\mathrm{d}x}\frac{\mathrm{d}}{\mathrm{d}x}\left[EI\frac{\mathrm{d}^2\upsilon}{\mathrm{d}x^2}\right] + wf \right) \mathrm{d}x + \left(w\frac{\mathrm{d}}{\mathrm{d}x}\left[EI\frac{\mathrm{d}^2\upsilon}{\mathrm{d}x^2}\right] \right)\Bigg|_0^L = 0. \tag{3.17}$$

Upon integrating by parts a second time, the following integro-differential equation is obtained:

$$\int_0^L \left(EI \frac{d^2 w}{dx^2} \frac{d^2 \upsilon}{dx^2} - wf \right) dx + \left(w \frac{d}{dx} \left[EI \frac{d^2 \upsilon}{dx^2} \right] - \frac{dw}{dx} \left[EI \frac{d^2 \upsilon}{dx^2} \right] \right) \Big|_0^L = 0. \quad (3.18)$$

This is the weak form of the governing ODE for Euler-Bernoulli beams. The complete weak form of the problem statement is the following. Find the deflection function $\upsilon(x)$ and its derivative $\theta(x) = -d\upsilon(x)/dx$ that satisfy (3.18), given beam properties $E(x)$, $I(x)$, and L, where $x \in [0,L]$, and subject to the boundary conditions at $x = 0, L$ and a distributed load function $f(x)$, a concentrated shearing force V_0 at $x = x_V$, a concentrated bending moment M_0 at $x = x_M$, an imposed rotation at $x = x_\theta$, and an imposed transverse displacement at $x = x_\upsilon$. Notice that (3.18) can be considered a second-order ODE in υ and in w.

Several remarks are in order that contrast beam theory with the 1-D bar/truss theory of Chapter 2. Specifically, we compare weak form ODEs (2.106) and (3.18). The former (bar/truss equation) involves a product of first derivatives of u and w; the latter (beam equation) involves a product of second derivatives of υ and w. The former involves a single set of two boundary conditions resulting from integration by parts once, while the latter involves two sets of two conditions stemming from integration by parts twice.

We will soon introduce shape functions in §3.3 and re-cast the problem domain to a single beam finite element. Then we shall see in §3.4 that the integral term involving the product of second derivatives of υ and w produces the finite element stiffness matrix, the integral term involving the distributed load produces the distributed load vector, the boundary term involving the the second derivative of deflection produces natural boundary conditions on bending moments, and the boundary term involving the third derivative of displacements produces natural boundary conditions on shearing forces.

To enable a general derivation of element equations, we introduce the sign conventions shown in Fig. 3.4 for degrees of freedom and their conjugate forces. Importantly, we note that positive shearing forces $Q_1^{(e)}$ and $Q_3^{(e)}$ always act in the $+y$-direction, and positive bending moments $Q_2^{(e)}$ and $Q_4^{(e)}$ always act in the clockwise sense. These conventions promote consistency regardless of from which direction an external observer views the beam.

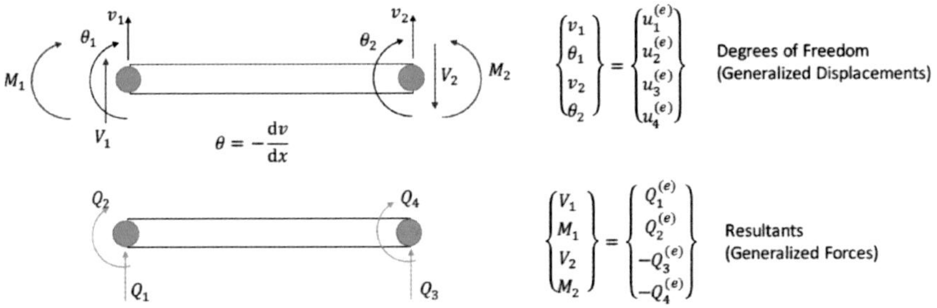

Fig. 3.4 Sign conventions for Euler-Bernoulli beam boundary conditions

3.3 Interpolation

The next step in derivation of the FEM for beams ensues introduction of shape functions for interpolation of primary field variables at locations differing from nodal points. Primary degrees of freedom for Euler-Bernoulli beams are the deflection $\upsilon(x)$ and slope $\theta(x) = -\mathrm{d}\upsilon(x)/\mathrm{d}x$. The selected shape functions must enable interpolation of both of these variables. For a beam element with two nodes, a polynomial of minimum order three is required to allow for interpolation given the deflection and slope at each node located at the endpoints of the beam. In other words, since a beam element has four degrees of freedom (the deflection and slope at each node), a cubic spline is minimally sufficient for accurate interpolation.

Polynomials that enable interpolation of a variable and its spatial derivative are called Hermite interpolation functions. Let an Euler-Bernoulli beam element occupy the domain $x \in [x_e, x_{e+1}]$, with length $L_e = x_{e+1} - x_e$. Applying the summation convention over repeated indices, the deflection in a single beam element $\upsilon^{(e)}$ is represented using the Hermite shape functions $N_i(x)$ as

$$\upsilon^{(e)}(x) = \sum_{i=1}^{4} N_i(x)u_i = N_i(x)u_i \qquad (x \in [x_e, x_{e+1}]). \tag{3.19}$$

The same form of equation is used to interpolate the slope $\theta^{(e)}$ in the element:

$$\theta^{(e)}(x) = -\frac{\mathrm{d}\upsilon^{(e)}(x)}{\mathrm{d}x} = -\sum_{i=1}^{4} \frac{\mathrm{d}N_i(x)}{\mathrm{d}x}u_i = -\frac{\mathrm{d}N_i(x)}{\mathrm{d}x}u_i \qquad (x \in [x_e, x_{e+1}]). \tag{3.20}$$

In these equations, u_i are components of the vector of generalized nodal displacements for the element:

$$\{\boldsymbol{u}\} = \begin{Bmatrix} u_1 \\ u_2 \\ u_3 \\ u_4 \end{Bmatrix} = \begin{Bmatrix} \upsilon_1 \\ \theta_1 \\ \upsilon_2 \\ \theta_2 \end{Bmatrix}, \tag{3.21}$$

and the deflection and slope at node j are (υ_j, θ_j), where $j = 1, 2$ in the present context.

Recall from §2.3.3 that shape functions must obey certain constraints, specifically N_i must have a value of 1 at the node for the shape function that corresponds to that node, and it must have a value of 0 at all other nodes. This constraint applies here for interpolation of deflection υ. For interpolation of slope θ, the negative derivative of N_i must have a value of 1 at the node for the shape function that corresponds to that node, and it must have a value of 0 at all other nodes. In equation form, these constraints correspond to

$$N_1(x_e) = 1, \quad N_2(x_e) = N_3(x_e) = N_4(x_e) = 0; \tag{3.22}$$

$$\frac{\mathrm{d}N_2(x_e)}{\mathrm{d}x} = -1, \quad \frac{\mathrm{d}N_1(x_e)}{\mathrm{d}x} = \frac{\mathrm{d}N_3(x_e)}{\mathrm{d}x} = \frac{\mathrm{d}N_4(x_e)}{\mathrm{d}x} = 0; \tag{3.23}$$

$$N_3(x_{e+1}) = 1, \quad N_1(x_{e+1}) = N_2(x_{e+1}) = N_4(x_{e+1}) = 0; \tag{3.24}$$

$$\frac{\mathrm{d}N_4(x_{e+1})}{\mathrm{d}x} = -1, \quad \frac{\mathrm{d}N_1(x_{e+1})}{\mathrm{d}x} = \frac{\mathrm{d}N_2(x_{e+1})}{\mathrm{d}x} = \frac{\mathrm{d}N_3(x_{e+1})}{\mathrm{d}x} = 0. \tag{3.25}$$

In order to simplify subsequent notation for interpolation and integration, and more importantly to enable the same general form of shape functions for all elements e, we now introduce the dimensionless natural coordinate ξ:

$$\xi = \frac{x - x_e}{L_e}, \qquad \xi \in [0,1]. \tag{3.26}$$

Also, from the chain rule of differentiation,

$$\frac{\mathrm{d}\xi}{\mathrm{d}x} = \frac{1}{L_e} \Rightarrow \mathrm{d}x = L_e \mathrm{d}\xi, \qquad \frac{\mathrm{d}N[(\xi(x)]}{\mathrm{d}x} = \frac{\mathrm{d}N}{\mathrm{d}\xi}\frac{\mathrm{d}\xi}{\mathrm{d}x} = \frac{1}{L_e}\frac{\mathrm{d}N}{\mathrm{d}\xi}; \tag{3.27}$$

$$\frac{\mathrm{d}^2N[(\xi(x)]}{\mathrm{d}x^2} = \frac{\mathrm{d}}{\mathrm{d}x}\left(\frac{\mathrm{d}N}{\mathrm{d}x}\right) = \frac{\mathrm{d}}{\mathrm{d}x}\left(\frac{1}{L_e}\frac{\mathrm{d}N}{\mathrm{d}\xi}\right) = \frac{\mathrm{d}}{\mathrm{d}\xi}\left(\frac{1}{L_e}\frac{\mathrm{d}N}{\mathrm{d}\xi}\right)\frac{\mathrm{d}\xi}{\mathrm{d}x} = \frac{1}{L_e^2}\frac{\mathrm{d}^2N}{\mathrm{d}\xi^2}. \tag{3.28}$$

The general form of the cubic Hermite shape functions is

$$N_i(x) = a_{i0} + a_{i1}x + a_{i2}x^2 + a_{i3}x^3. \tag{3.29}$$

Coefficients $a_{ij}\,(i = 1, \ldots, 4; j = 0, \ldots, 3)$ can be determined using the requisite conditions listed in (3.22)–(3.25). The resulting shape functions in terms of ξ are as follows:

$$N_1(\xi) = 1 - 3\xi^2 + 2\xi^3, \tag{3.30}$$

$$N_2(\xi) = -L_e\xi(1-\xi)^2, \tag{3.31}$$

$$N_3(\xi) = 3\xi^2 - 2\xi^3, \tag{3.32}$$

$$N_4(\xi) = -L_e\xi(\xi^2 - \xi). \tag{3.33}$$

These functions and their derivatives $N_i' = \mathrm{d}N_i/\mathrm{d}x$ are respectively shown on the left and right sides of Fig. 3.5.

In matrix form, the shape functions occupy the row vector $\boldsymbol{N}$:

$$\begin{aligned}[\boldsymbol{N}] &= [N_1 \quad N_2 \quad N_3 \quad N_4] \\ &= \left[1 + \xi^2(2\xi - 3) \quad -L_e\xi(1-\xi)^2 \quad \xi^2(3-2\xi) \quad L_e\xi^2(1-\xi)\right].\end{aligned} \tag{3.34}$$

In order to compute the finite element stiffness matrix, the second derivatives of $N_i(x)$ will be needed. Using (3.28), these are determined to be

$$\frac{\mathrm{d}^2N_1}{\mathrm{d}x^2} = \frac{6}{L_e^2}(2\xi - 1), \tag{3.35}$$

$$\frac{\mathrm{d}^2N_2}{\mathrm{d}x^2} = \frac{2}{L_e}(2 - 3\xi), \tag{3.36}$$

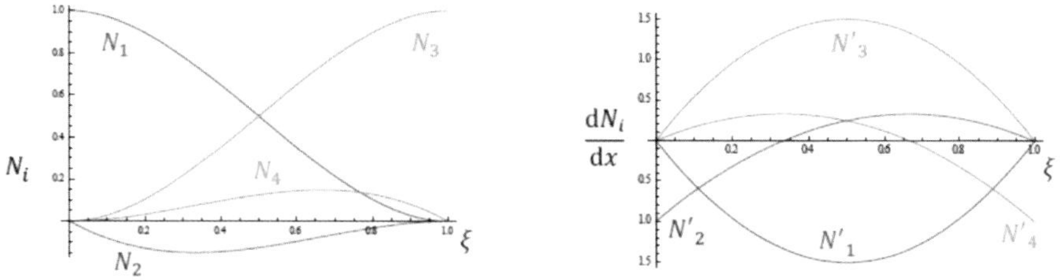

Fig. 3.5 Beam shape functions (left) and their first derivatives (right)

$$\frac{\mathrm{d}^2 N_3}{\mathrm{d}x^2} = \frac{6}{L_e^2}(1-2\xi), \tag{3.37}$$

$$\frac{\mathrm{d}^2 N_4}{\mathrm{d}x^2} = \frac{2}{L_e}(1-3\xi). \tag{3.38}$$

In matrix form, the second derivatives occupy the "B-matrix", i.e., the row vector $\boldsymbol{B}$ that is used in calculation of the element stiffness matrix:

$$\begin{aligned}[\boldsymbol{B}] &= \begin{bmatrix} \frac{\mathrm{d}^2 N_1}{\mathrm{d}x^2} & \frac{\mathrm{d}^2 N_2}{\mathrm{d}x^2} & \frac{\mathrm{d}^2 N_3}{\mathrm{d}x^2} & \frac{\mathrm{d}^2 N_4}{\mathrm{d}x^2} \end{bmatrix} \\ &= \begin{bmatrix} \frac{6}{L_e^2}(2\xi-1) & \frac{2}{L_e}(2-3\xi) & \frac{6}{L_e^2}(1-2\xi) & \frac{2}{L_e}(1-3\xi) \end{bmatrix}. \end{aligned} \tag{3.39}$$

3.4 Matrix Equations

Returning now to derivation of the beam element equilibrium equations, consider again the weak form of the governing differential equation in (3.18), repeated below for an element with nodes at (x_e, x_{e+1}):

$$\int_{x_e}^{x_{e+1}} \left(EI \frac{\mathrm{d}^2 w}{\mathrm{d}x^2} \frac{\mathrm{d}^2 \upsilon}{\mathrm{d}x^2} - wf \right) \mathrm{d}x + \left(w \frac{\mathrm{d}}{\mathrm{d}x} \left[EI \frac{\mathrm{d}^2 \upsilon}{\mathrm{d}x^2} \right] - \frac{\mathrm{d}w}{\mathrm{d}x} \left[EI \frac{\mathrm{d}^2 \upsilon}{\mathrm{d}x^2} \right] \right) \Big|_{x_e}^{x_{e+1}} = 0. \tag{3.40}$$

Next, substituting from (3.19) for the deflection $\upsilon(x)$ in terms of shape functions $N_i(x)$ and nodal degrees of freedom u_i, and making a similar substitution for the weight function $w(x)$ in terms of the same shape functions and analogous nodal values of w and $-\mathrm{d}w/\mathrm{d}x$ written as $w_j\,(j = 1, \ldots, 4)$, i.e.,

$$\begin{Bmatrix} w_1 \\ w_2 \\ w_3 \\ w_4 \end{Bmatrix} = \begin{Bmatrix} w(x_e) \\ -\frac{\mathrm{d}w}{\mathrm{d}x}(x_e) \\ w(x_{e+1}) \\ -\frac{\mathrm{d}w}{\mathrm{d}x}(x_{e+1}) \end{Bmatrix}, \qquad w(x) = N_j(x)w_j; \tag{3.41}$$

the weak form equation can be written as

$$\begin{aligned} &\int_{x_e}^{x_{e+1}} EI\frac{\mathrm{d}^2N_i}{\mathrm{d}x^2}\frac{\mathrm{d}^2N_j}{\mathrm{d}x^2}u_iw_j\mathrm{d}x - \int_{x_e}^{x_{e+1}} N_jw_jf\mathrm{d}x \\ &\quad + \left(N_jw_j\frac{\mathrm{d}}{\mathrm{d}x}\left[EI\frac{\mathrm{d}^2\upsilon}{\mathrm{d}x^2}\right] - w_j\frac{\mathrm{d}N_j}{\mathrm{d}x}\left[EI\frac{\mathrm{d}^2\upsilon}{\mathrm{d}x^2}\right]\right)\Bigg|_{x_e}^{x_{e+1}} = 0. \end{aligned} \tag{3.42}$$

Since each u_i and w_j is a constant for a given element, these terms can be moved outside the integral operators. Furthermore, since $w(x)$ and thus each w_j is arbitrary, (3.42) implies that the following system of four integro-differential equations must hold (one equation for each value of $j = 1,\ldots,4$):

$$\begin{aligned} &\left[\int_{x_e}^{x_{e+1}} EI\frac{\mathrm{d}^2N_i}{\mathrm{d}x^2}\frac{\mathrm{d}^2N_j}{\mathrm{d}x^2}\mathrm{d}x\right]u_i - \int_{x_e}^{x_{e+1}} N_jf\mathrm{d}x \\ &\quad + \left(N_j\frac{\mathrm{d}}{\mathrm{d}x}\left[EI\frac{\mathrm{d}^2\upsilon}{\mathrm{d}x^2}\right] - \frac{\mathrm{d}N_j}{\mathrm{d}x}\left[EI\frac{\mathrm{d}^2\upsilon}{\mathrm{d}x^2}\right]\right)\Bigg|_{x_e}^{x_{e+1}} = 0. \end{aligned} \tag{3.43}$$

The term in square braces on the left side of (3.43) will become the element stiffness matrix $\boldsymbol{k}^{(e)}$. Noting that the second derivatives of the shape functions comprise $\boldsymbol{B}$ of (3.39),

$$k_{ij}^{(e)} = \int_{x_e}^{x_{e+1}} EI\frac{\mathrm{d}^2N_i}{\mathrm{d}x^2}\frac{\mathrm{d}^2N_j}{\mathrm{d}x^2}\mathrm{d}x = \int_{x_e}^{x_{e+1}} EIB_iB_j\mathrm{d}x, \tag{3.44}$$

which can be written in vector-matrix form in terms of natural coordinates via (3.27) as

$$[\boldsymbol{k}^{(e)}] = \int_0^1 EI[\boldsymbol{B}]^{\mathrm{T}}[\boldsymbol{B}]L_e\mathrm{d}\xi. \tag{3.45}$$

When EI is a constant over the element (i.e., when this product does not depend on x or ξ), the end result of the integration in (3.45), with use of (3.39), is the following 4×4 element stiffness matrix:

$$[\boldsymbol{k}^{(e)}] = \frac{2EI}{L_e^3}\begin{bmatrix} 6 & -3L_e & -6 & -3L_e \\ -3L_e & 2L_e^2 & 3L_e & L_e^2 \\ -6 & 3L_e & 6 & 3L_e \\ -3L_e & L_e^2 & 3L_e & 2L_e^2 \end{bmatrix}. \tag{3.46}$$

The second integral term on the left side of (3.43) is the distributed load vector $\boldsymbol{f}^{(e)}$. Transforming the domain to natural coordinates and invoking vector-matrix notation, this vector is

$$\{\boldsymbol{f}^{(e)}\} = \int_0^1 [\boldsymbol{N}]^{\mathrm{T}} f(\xi)L_e\mathrm{d}\xi, \tag{3.47}$$

where the shape function matrix N is the row vector in (3.34). For dimensional consistency, $f(\xi)$, like $f(x)$, has units of force per unit length. For the special case that f is equal to a constant f_0 over the element, the integral is

$$\{f^{(e)}\} = \frac{f_0 L_e}{12}\begin{Bmatrix} 6 \\ -L_e \\ 6 \\ L_e \end{Bmatrix}. \qquad (f = f_0 = \text{constant}) \tag{3.48}$$

When the product $N(\xi)\cdot f(\xi)$ is difficult or impossible to integrate analytically, a cubic interpolation of f may be invoked via the Hermite shape functions:

$$f(\xi) = N_i(\xi)\hat{f}_i^{(e)}, \qquad \{\hat{f}^{(e)}\} = \begin{Bmatrix} \hat{f}_1^{(e)} \\ \hat{f}_2^{(e)} \\ \hat{f}_3^{(e)} \\ \hat{f}_4^{(e)} \end{Bmatrix} = \begin{Bmatrix} f(\xi = 0) \\ 0 \\ f(\xi = 1) \\ 0 \end{Bmatrix}. \tag{3.49}$$

Substituting (3.49) back into (3.47) then gives the approximate distributed load vector in terms of spatially constant nodal values $\hat{f}_i^{(e)}$:

$$\{f^{(e)}\} = \int_0^1 [N]^{\mathrm{T}}[N]\{\hat{f}^{(e)}\}L_e \mathrm{d}\xi = \left[\int_0^1 [N]^{\mathrm{T}}[N]L_e \mathrm{d}\xi\right]\{\hat{f}^{(e)}\}, \tag{3.50}$$

which can then be evaluated analytically regardless of the complexity of the original distributed load function $f[\xi(x)]$.

The remaining terms in equilibrium equation (3.43) correspond to natural boundary conditions applied at the nodes or endpoints of the beam ($x = x_e, x_{e+1} \leftrightarrow \xi = 0, 1$). The following definitions are consistent with applied nodal loads shown in Fig. 3.4:

$$Q_1^{(e)} = \left[\frac{\mathrm{d}}{\mathrm{d}x}\left(EI\frac{\mathrm{d}^2 \upsilon}{\mathrm{d}x^2}\right)\right]_{x_e}, \qquad Q_2^{(e)} = \left[EI\frac{\mathrm{d}^2 \upsilon}{\mathrm{d}x^2}\right]_{x_e}; \tag{3.51}$$

$$Q_3^{(e)} = -\left[\frac{\mathrm{d}}{\mathrm{d}x}\left(EI\frac{\mathrm{d}^2 \upsilon}{\mathrm{d}x^2}\right)\right]_{x_{e+1}}, \qquad Q_4^{(e)} = -\left[EI\frac{\mathrm{d}^2 \upsilon}{\mathrm{d}x^2}\right]_{x_{e+1}}. \tag{3.52}$$

Recall that $Q_1^{(e)}$ and $Q_3^{(e)}$ are transverse forces, and that $Q_2^{(e)}$ and $Q_4^{(e)}$ are bending moments. Substituting definitions above into the aforementioned remaining terms in (3.43) gives

$$\begin{aligned} &\left[N_j\frac{\mathrm{d}}{\mathrm{d}x}\left(EI\frac{\mathrm{d}^2 \upsilon}{\mathrm{d}x^2}\right) - \frac{\mathrm{d}N_j}{\mathrm{d}x}\left(EI\frac{\mathrm{d}^2 \upsilon}{\mathrm{d}x^2}\right)\right]\Bigg|_{x_e}^{x_{e+1}} \\ &= -N_j(x_e)Q_1^{(e)} + \frac{\mathrm{d}N_j}{\mathrm{d}x}(x_e)Q_2^{(e)} - N_j(x_{e+1})Q_3^{(e)} + \frac{\mathrm{d}N_j}{\mathrm{d}x}(x_{e+1})Q_4^{(e)}. \end{aligned} \tag{3.53}$$

Moving these terms to the opposite side of the equation and using (3.22)–(3.25), the four terms in (3.53) can be shown to simply comprise the column vector of nodal forces for the beam element:

$$N_j(x_e)Q_1^{(e)} - \frac{dN_j}{dx}(x_e)Q_2^{(e)} + N_j(x_{e+1})Q_3^{(e)} - \frac{dN_j}{dx}(x_{e+1})Q_4^{(e)} \rightarrow \begin{Bmatrix} Q_1^{(e)} \\ Q_2^{(e)} \\ Q_3^{(e)} \\ Q_4^{(e)} \end{Bmatrix}. \tag{3.54}$$

Finally, combining (3.21), (3.46), (3.47), and (3.54), the weak form expression (3.43) can be written in vector-matrix notation for an individual Euler-Bernoulli beam element e with constant EI as

$$\frac{2(EI)^{(e)}}{L_e^3}\begin{bmatrix} 6 & -3L_e & -6 & -3L_e \\ -3L_e & 2L_e^2 & 3L_e & L_e^2 \\ -6 & 3L_e & 6 & 3L_e \\ -3L_e & L_e^2 & 3L_e & 2L_e^2 \end{bmatrix}\begin{Bmatrix} \upsilon_1^{(e)} \\ \theta_1^{(e)} \\ \upsilon_2^{(e)} \\ \theta_2^{(e)} \end{Bmatrix} = \begin{Bmatrix} Q_1^{(e)} \\ Q_2^{(e)} \\ Q_3^{(e)} \\ Q_4^{(e)} \end{Bmatrix} + \begin{Bmatrix} f_1^{(e)} \\ f_2^{(e)} \\ f_3^{(e)} \\ f_4^{(e)} \end{Bmatrix} = \begin{Bmatrix} F_1^{(e)} \\ F_2^{(e)} \\ F_3^{(e)} \\ F_4^{(e)} \end{Bmatrix}. \tag{3.55}$$

The total load vector accounting for forces and moments applied directly to the nodes ($\boldsymbol{Q}^{(e)}$) and distributed loads resolved at the nodes ($\boldsymbol{f}^{(e)}$) is denoted by the vector on the far right, $\boldsymbol{F}^{(e)}$.

3.5 Assembly, Solution, and Post-processing

The assembly of a global stiffness matrix and a global force vector for two beam elements in series is depicted graphically in Fig. 3.6. The same notational conventions introduced in Chapter 2 in the context of truss elements regarding superscripts are followed here in the context of beam elements: superscripts denote element numbers. Occasionally, as in Fig. 3.6, parentheses enclosing such superscripts are omitted when there is no chance for confusion. When superscripts are present, subscripts denote local degree-of-freedom numbers which run from 1 to 4 for Euler-Bernoulli beam elements.

For the example shown in Fig. 3.6, no distributed loads are applied, so $\boldsymbol{F}^{(e)} = \boldsymbol{Q}^{(e)}$. Furthermore, no concentrated shearing forces or bending moments are applied at the shared node where the elements connect (node 2 of element 1 = node 1 of element 2 = global node 2), such that the following natural boundary conditions hold:

$$F_3 = Q_3^{(1)} + Q_1^{(2)} = 0, \qquad F_4 = Q_4^{(1)} + Q_2^{(2)} = 0. \tag{3.56}$$

The connectivity of the structure requires that the deflection and rotation are the same at the right end of element 1 and the left end of element 2:

$$u_3 = \upsilon_2^{(1)} = u_3^{(1)} = \upsilon_1^{(2)} = u_1^{(2)}, \qquad u_4 = \theta_2^{(1)} = u_4^{(1)} = \theta_1^{(2)} = u_2^{(2)}. \tag{3.57}$$

Each beam element supplies four local equilibrium equations of the form in (3.55), for a total of eight equations among the two elements. Using the two independent equations in the conditions (3.57), this system can be reduced to six equations in six unknown global displacements u_i $(i = 1,\ldots,6)$, recalling that general displacement vector $\boldsymbol{u}$ includes both nodal deflections and nodal rotations. In vector-matrix notation, the global system is of the usual form

$$[\boldsymbol{K}]\{\boldsymbol{u}\} = \{\boldsymbol{F}\}, \tag{3.58}$$

where here the global stiffness $\boldsymbol{K}$ is the 6×6 assembled matrix shown in Fig. 3.6, and the assembled force vector $\boldsymbol{F}$ is the sum of corresponding nodal load(s) $\boldsymbol{Q}^{(e)}$ where nodes are shared. Empty entries in Fig. 3.6 in the symmetric matrix $\boldsymbol{K}$ are zeroes. Details of the assembly process for two elements can be verified from examination of the individual algebraic equations. Such an exercise, albeit tedious, is left to the reader. All modern FEA software performs the assembly of global stiffness matrices and force vectors for the user.

Boundary conditions are invoked on the global structure following the same general rules introduced in Chapter 2 (e.g., equivalently §2.2 or §2.3). Essential boundary conditions on nodal deflection υ or nodal rotation θ are invoked mathematically by zeroing corresponding rows and columns of the global stiffness matrix and moving any necessary contributions (if nonzero deflections or rotations are imposed) to the right side of the global system, such that $\boldsymbol{F}$ in (3.58) contains such contributions. Natural boundary conditions on shear force V or bending moment M are invoked directly as entries in the global force vector $\boldsymbol{Q}$, recalling that a node with no applied forces or constraints would have a zero entry in the corresponding component of force. Distributed loads also contribute to $\boldsymbol{F}$ as manifested as local nodal forces via (3.55). Once boundary conditions have been imposed, the global system should be statically determinate, the condensed stiffness $\boldsymbol{K}$ can be inverted, and an equation of the form in (3.58) can be solved for primary variables u_i:

$$\{\boldsymbol{u}\} = [\boldsymbol{K}]^{-1}\{\boldsymbol{F}\}. \tag{3.59}$$

The primary solution $\boldsymbol{u}$ contains nodal deflections and rotations referred to global degree-of-freedom numbers. As is the case for assembly, any modern FEM software performs the mathematical operations associated with boundary conditions (i.e., once boundary conditions are specified by the user, the code condenses the global system accordingly), stiffness matrix inversion, and solution of the algebraic system (3.59).

Post-processing for beam elements is now considered briefly. Recall from (3.19) and (3.20) that generalized displacements u_i are the primary solution variables, consisting of nodal deflections $(u_1^{(e)}, u_3^{(e)})$ and nodal rotations $(u_2^{(e)}, u_4^{(e)})$. Secondary variables at the nodes are the reaction shearing forces corresponding to generalized forces $(Q_1^{(e)}, Q_3^{(e)})$ and the reaction bending moments corresponding to generalized forces $(Q_2^{(e)}, Q_4^{(e)})$. Such reaction forces and moments can be calculated by returning to the element equilibrium equations of (3.55) and back-substituting the primary

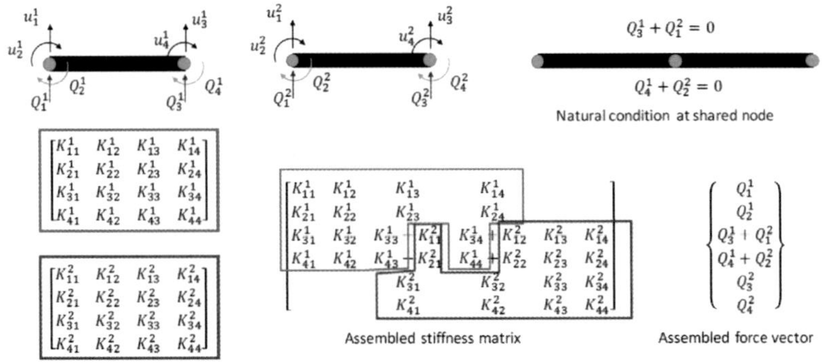

Fig. 3.6 Assembly of global matrix equations for a system of two serial beam elements

solution variables which have now been determined from (3.59):

$$\begin{Bmatrix} Q_1^{(e)} \\ Q_2^{(e)} \\ Q_3^{(e)} \\ Q_4^{(e)} \end{Bmatrix} = \frac{2(EI)^{(e)}}{L_e^3} \begin{bmatrix} 6 & -3L_e & -6 & -3L_e \\ -3L_e & 2L_e^2 & 3L_e & L_e^2 \\ -6 & 3L_e & 6 & 3L_e \\ -3L_e & L_e^2 & 3L_e & 2L_e^2 \end{bmatrix} \begin{Bmatrix} u_1^{(e)} \\ u_2^{(e)} \\ u_3^{(e)} \\ u_4^{(e)} \end{Bmatrix} - \begin{Bmatrix} f_1^{(e)} \\ f_2^{(e)} \\ f_3^{(e)} \\ f_4^{(e)} \end{Bmatrix}. \tag{3.60}$$

Another quantity frequently sought during post-processing is the bending stress $\sigma(x,y)$ at a point $x \in (x_e, x_{e+1})$, where y is the vertical distance from the neutral axis. To determine the stress, the bending moment $M(x)$ is calculated first using (3.7) in conjunction with interpolation via (3.19) and the definition of $\boldsymbol{B}$ in terms of second derivatives in (3.39):

$$M(x) = E(x)I(x)\frac{d^2 \upsilon}{dx^2}(x) = E(x)I(x)\frac{d^2}{dx^2}\left([\boldsymbol{N}(x)]\{u^{(e)}\}\right) = E(x)I(x)[\boldsymbol{B}(x)]\{u^{(e)}\}. \tag{3.61}$$

The stress at a point with coordinates (x,y) is then found via manipulation of (3.5), (3.6), and (3.7):

$$\sigma(x,y) = E(x)\varepsilon(x,y) = -E(x)y\frac{d^2\upsilon(x)}{dx^2} = -\frac{M(x)y}{I(x)}. \tag{3.62}$$

Substitution of (3.61) into (3.62) then gives the final resulting stress in terms of generalized nodal displacements for element e:

$$\sigma(x,y) = -E(x)y[\boldsymbol{B}(x)]\{\boldsymbol{u}^{(e)}\}. \tag{3.63}$$

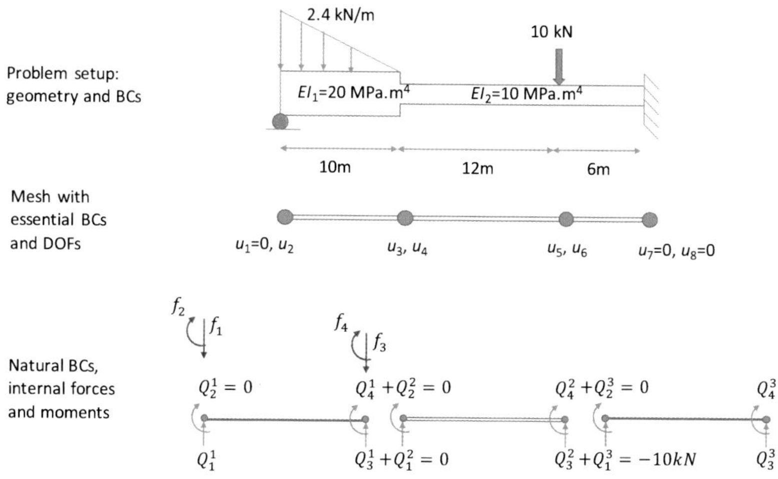

Fig. 3.7 Beam example problem 1: determine the primary solution (all unknown degrees of freedom) as well as nodal forces and stress in element 2 (the 12 m section), the latter at node 2

3.6 Beam Example Problems

Three representative example problems involving Euler-Bernoulli beams are considered next.

3.6.1 *Example 1*

The first example is given in Fig. 3.7, with the physical problem shown at the top of the image. A beam is comprised of two sections with different values of EI as shown. The left end of the beam, taken at $x = 0$, is prevented from deflecting but is free to rotate. The right end of the beam, at $x = 28$ m, is fully clamped and thus prevented from deflecting and rotating. A distributed load acting downward is applied to the left section, decreasing linearly in magnitude with increasing x. The objectives, as listed in the figure caption, are determination of the primary solution and the forces and stress in the midsection of the beam at $x = 12$ m.

The solution is obtained as follows. The generalized global displacement vector is

$$\{\boldsymbol{u}\} = [u_1 \quad u_2 \quad u_3 \quad u_4 \quad u_5 \quad u_6 \quad u_7 \quad u_8]^{\mathrm{T}}, \tag{3.64}$$

where entries of $\boldsymbol{u}$ with odd subscripts are deflections and entries with even subscripts are rotations. As is clear from the middle diagram in Fig. 3.7, $u_1 = u_7 = u_8 = 0$. Natural boundary conditions, internal forces, and internal moments are shown in the bottom diagram of Fig. 3.7. Recall that the internal forces $Q_i^{(e)}$ sum to zero when no corresponding applied force is given. Contributions of the distributed load resolved at nodes 1 and 2 are quantified by f_i, where $i = 1, \ldots, 4$. Specifically applying (3.47), these contributions are

$$\begin{aligned}\int_0^1 [N]^{\mathrm{T}} f(\xi) L_e \mathrm{d}\xi &= -24\ \mathrm{kN} \int_0^1 [N]^{\mathrm{T}} (1-\xi) \mathrm{d}\xi = [f_1 \quad f_2 \quad f_3 \quad f_4]^{\mathrm{T}} \\ &= [-8.4\ \mathrm{kN} \quad 12\ \mathrm{kN\ m} \quad -3.6\ \mathrm{kN} \quad -8\ \mathrm{kN\ m}]^{\mathrm{T}}.\end{aligned} \tag{3.65}$$

The global stiffness matrix $\boldsymbol{K}$ is obtained by assembling the stiffness matrices of the three beam elements in series, where node 2 of element 1 = node 1 of element 2 and node 2 of element 2 = node 1 of element 3. Using (3.46), which is relevant here since each element has constant product EI, the global stiffness matrix is then assembled numerically in consistent force units of kN as the following symmetric 8×8 matrix:

$$[\boldsymbol{K}] = \begin{bmatrix} 240 & -1200 & -240 & -1200 & 0 & 0 & 0 & 0 \\ -1200 & 8000 & 1200 & 4000 & 0 & 0 & 0 & 0 \\ -240 & 1200 & 309 & 783 & -69 & -417 & 0 & 0 \\ -1200 & 4000 & 783 & 11333 & 417 & 1667 & 0 & 0 \\ 0 & 0 & -69 & 417 & 625 & -1250 & -556 & -1667 \\ 0 & 0 & -417 & 1667 & -1250 & 10000 & 1667 & 3333 \\ 0 & 0 & 0 & 0 & -556 & 1667 & 556 & 1667 \\ 0 & 0 & 0 & 0 & -1667 & 3333 & 1667 & 6667 \end{bmatrix}. \tag{3.66}$$

Inserting (3.65) and the natural boundary conditions shown in the bottom diagram of Fig. 3.7, the assembled global force vector in force units of kN is

$$\{\boldsymbol{F}\} = \begin{Bmatrix} f_1 \\ f_2 \\ f_3 \\ f_4 \\ 0 \\ 0 \\ 0 \\ 0 \end{Bmatrix} + \begin{Bmatrix} Q_1^1 \\ Q_2^1 \\ Q_3^1 + Q_1^2 \\ Q_4^1 + Q_2^2 \\ Q_3^2 + Q_1^3 \\ Q_4^2 + Q_2^3 \\ Q_3^3 \\ Q_4^3 \end{Bmatrix} = \begin{Bmatrix} -8.4 \\ 12 \\ -3.6 \\ -8 \\ 0 \\ 0 \\ 0 \\ 0 \end{Bmatrix} + \begin{Bmatrix} Q_1^1 \\ 0 \\ 0 \\ 0 \\ -10 \\ 0 \\ Q_3^3 \\ Q_4^3 \end{Bmatrix}. \tag{3.67}$$

Applying the essential boundary conditions $u_1 = u_7 = u_8 = 0$, rows/columns 1, 7, and 8 can be eliminated from the global system of equations, leading to the condensed system of five equations and five primary unknowns u_2, u_3, u_4, u_5, u_6:

$$\begin{bmatrix} 8000 & 1200 & 4000 & 0 & 0 \\ 1200 & 309 & 783 & -69 & -417 \\ 4000 & 783 & 11333 & 417 & 1667 \\ 0 & -69 & 417 & 625 & -1250 \\ 0 & -417 & 1667 & -1250 & 10000 \end{bmatrix} \begin{Bmatrix} u_2 \\ u_3 \\ u_4 \\ u_5 \\ u_6 \end{Bmatrix} = \begin{Bmatrix} 12 \\ -3.6 \\ -8 \\ -10 \\ 0 \end{Bmatrix}. \quad (3.68)$$

Inverting the condensed stiffness matrix numerically and solving for the primary variables gives the primary solution:

$$\begin{Bmatrix} u_2 \\ u_3 \\ u_4 \\ u_5 \\ u_6 \end{Bmatrix} = \begin{Bmatrix} 0.018 \text{ rad} \\ -0.133 \text{ m} \\ 0.007 \text{ rad} \\ -0.065 \text{ m} \\ -0.015 \text{ rad} \end{Bmatrix}. \quad (3.69)$$

With the primary solution available, post-processing is now possible. Internal nodal forces in element 2 are obtained by substituting values of $u_1^2 = u_3$, $u_2^2 = u_4$, $u_3^2 = u_5$, and $u_4^2 = u_6$ into an element equilibrium equation of the form in (3.55), giving

$$\{\boldsymbol{Q}^{(2)}\} = [\boldsymbol{k}^{(2)}]\{\boldsymbol{u}^{(2)}\} = \begin{Bmatrix} Q_1^2 \\ Q_2^2 \\ Q_3^2 \\ Q_4^2 \end{Bmatrix} = \begin{Bmatrix} -1.38 \text{ kN} \\ 26.16 \text{ kN m} \\ 1.38 \text{ kN} \\ -9.54 \text{ kN m} \end{Bmatrix}. \quad (3.70)$$

Finally, we compute the stress in global node 2, which is local node 1 of element 2, located at $x = 12$ m or $\xi = 0$ for element 2. Using (3.63) with the B-matrix components determined from (3.39),

$$\sigma(y)|_{\xi=0} = -Ey[\boldsymbol{B}]_{\xi=0} \begin{Bmatrix} u_1^2 \\ u_2^2 \\ u_3^2 \\ u_4^2 \end{Bmatrix} = -2.62 \times Ey \quad (E \text{ in kPa; } y \text{ in m}). \quad (3.71)$$

3.6.2 Example 2

The second example problem is shown in Fig. 3.8 and is adapted from [2]. An Euler-Bernoulli beam is fixed to a wall at the left end at $x = 0$ and is supported transversely by a linear spring with stiffness constant k at the right end at $x = 10$ ft. A downward oriented distributed load is also applied, increasing in magnitude from zero to 100 lb/ft as x increases from 4 ft to 10 ft. The entire beam has a constant product of elastic modulus and moment of inertia EI.

The geometry and loading are such that two beam elements in series are sufficient, with element 2 supporting the distributed load. The objective, as stated in the caption of the figure, is determination of the primary solution. As depicted in the upper diagram of Fig. 3.8, the left end is fixed or fully clamped; essential boundary conditions corresponding to this fixed end at $x = 0$ are

$$\upsilon(0\,\text{m}) = u_1 = 0, \qquad \theta(0\,\text{m}) = -(\text{d}\upsilon/\text{d}x)(0\,\text{m}) = u_2 = 0. \tag{3.72}$$

The primary unknowns are u_3, u_4, u_5, and u_6 where subscripts denote global degree-of-freedom numbers. Natural boundary conditions at the center node from null concentrated forces lead to

$$\frac{\text{d}}{\text{d}x}\left(EI\frac{\text{d}^2\upsilon}{\text{d}x^2}\right)(4\,\text{m}) = 0 = Q_3^1 + Q_1^2, \qquad \left(EI\frac{\text{d}^2\upsilon}{\text{d}x^2}\right)(4\,\text{m}) = 0 = Q_4^1 + Q_2^2. \tag{3.73}$$

At the right end, no moment is applied, but the spring induces a transverse shearing force proportional to and with a sign opposite to those of the vertical endpoint displacement u_5:

$$-\frac{\text{d}}{\text{d}x}\left(EI\frac{\text{d}^2\upsilon}{\text{d}x^2}\right)(10\,\text{m}) = -ku_5 = Q_3^2, \qquad -\left(EI\frac{\text{d}^2\upsilon}{\text{d}x^2}\right)(10\,\text{m}) = 0 = Q_4^2. \tag{3.74}$$

The contribution from the distributed load produces effective shearing forces and bending moments at global nodes 2 and 3 belonging to element 2. These generalized forces are computed as

$$\begin{aligned}\{\boldsymbol{f}^{(2)}\} &= \int_0^1 [\boldsymbol{N}]^{\text{T}} f(\xi) L_e \text{d}\xi = -600\,\text{lb} \int_0^1 [\boldsymbol{N}]^{\text{T}} \xi \text{d}\xi \\ &= [-90\,\text{lb} \quad 120\,\text{lb ft} \quad -210\,\text{lb} \quad -180\,\text{lb ft}]^{\text{T}}.\end{aligned} \tag{3.75}$$

The assembled total force vector $\boldsymbol{F} = \boldsymbol{Q} + \boldsymbol{f}$ is thus

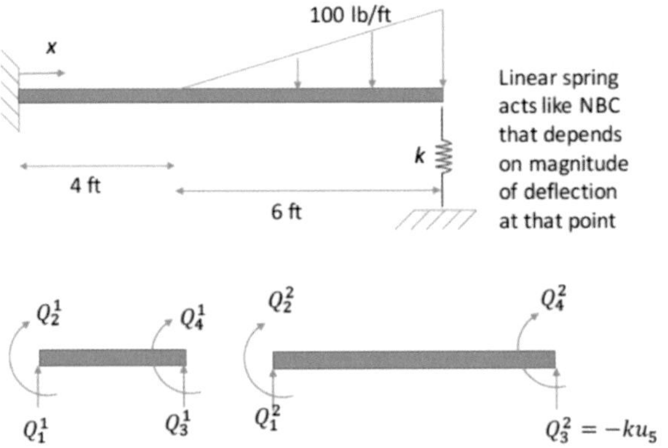

Fig. 3.8 Beam example problem 2: determine the primary solution in symbolic form

$$\{\boldsymbol{F}\} = \begin{Bmatrix} Q_1^1 \\ Q_2^1 \\ Q_3^1 + Q_1^2 + f_1^2 \\ Q_4^1 + Q_2^2 + f_2^2 \\ Q_3^2 + f_3^2 \\ Q_4^2 + f_4^2 \end{Bmatrix} = \begin{Bmatrix} Q_1^1 \\ Q_2^1 \\ -90 \text{ lb} \\ 120 \text{ lb ft} \\ -210 - ku_5 \text{ lb} \\ -180 \text{ lb ft} \end{Bmatrix}. \tag{3.76}$$

Consulting (3.46), stiffness matrices for elements 1 and 2 are computed, respectively, with consistent base units of ft and lb, as

$$[\boldsymbol{k}^{(1)}] = \frac{EI}{144}\begin{bmatrix} 27 & -54 & -27 & -54 \\ -54 & 144 & 54 & 72 \\ -27 & 54 & 27 & 54 \\ -54 & 72 & 54 & 144 \end{bmatrix}, \tag{3.77}$$

$$[\boldsymbol{k}^{(2)}] = \frac{EI}{144}\begin{bmatrix} 8 & -24 & -8 & -24 \\ -24 & 96 & 24 & 48 \\ -8 & 24 & 8 & 24 \\ -24 & 48 & 24 & 96 \end{bmatrix}. \tag{3.78}$$

Referring to Fig. 3.6, the assembled stiffness matrix for the system is then the symmetric 6×6 matrix

$$[\boldsymbol{K}] = \bigwedge_{e=1,2} [\boldsymbol{k}^{(e)}] = \frac{EI}{144}\begin{bmatrix} 27 & -54 & -27 & -54 & 0 & 0 \\ -54 & 144 & 54 & 72 & 0 & 0 \\ -27 & 54 & 35 & 30 & -8 & -24 \\ -54 & 72 & 30 & 240 & 24 & 48 \\ 0 & 0 & -8 & 24 & 8 & 24 \\ 0 & 0 & -24 & 48 & 24 & 96 \end{bmatrix}. \tag{3.79}$$

Prior to application of boundary conditions, the global system $\boldsymbol{Ku} = \boldsymbol{F}$ consists of six algebraic equations and is statically indeterminate. Application of essential boundary conditions (3.72) enables removal of the first two equations (e.g., rows/columns 1 and 2 of the original stiffness matrix $\boldsymbol{K}$). Then moving the contribution of the spring to the left side, the global system in condensed form consists of four equations in four unknown displacements:

$$\frac{EI}{144}\begin{bmatrix} 35 & 30 & -8 & -24 \\ 30 & 240 & 24 & 48 \\ -8 & 24 & 8 + (144k)/(EI) & 24 \\ -24 & 48 & 24 & 96 \end{bmatrix} \begin{Bmatrix} u_3 \\ u_4 \\ u_5 \\ u_6 \end{Bmatrix} = \begin{Bmatrix} -90 \text{ lb} \\ 120 \text{ lb ft} \\ -210 \text{ lb} \\ -180 \text{ lb ft} \end{Bmatrix}. \tag{3.80}$$

The primary solution (u_3, u_4, u_5, u_6) can be obtained by inverting the condensed 4×4 stiffness matrix on the left side once numerical values of geometric parameter I and constitutive parameters E and k are prescribed.

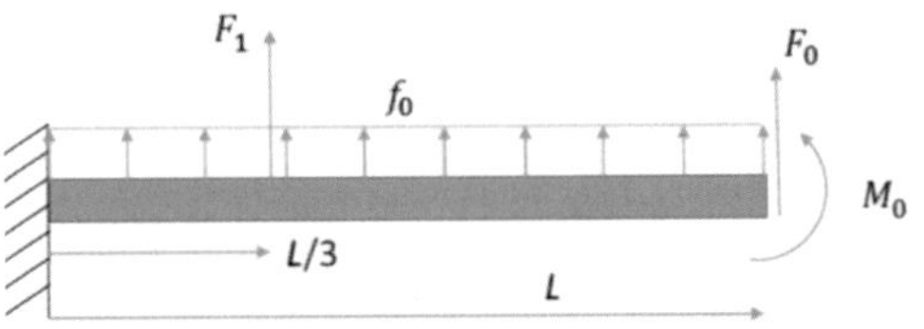

Fig. 3.9 Beam example problem 3: determine the total force vector for this beam when represented as a single finite element

3.6.3 Example 3

The third and final beam example problem is depicted in Fig. 3.9. The objective, as stated in the figure caption, is to write the total force vector $\boldsymbol{F}$ for this beam, assuming the whole beam is represented via a single finite element. The loads applied are a concentrated shearing force F_0 and concentrated bending moment M_0 at the right end, a constant distributed load of magnitude f_0, and a point force F_1 applied one third of the distance along the total length L. The beam is fully clamped or fixed at its left end, at $x = 0$.

First consider reaction forces at the left end, where the beam is rigidly fixed to the wall. Since deflection and rotation are both imposed as zero at $x = 0$, the reaction forces Q_1 and Q_2 are unknowns to be determined as an outcome of post-processing. Next consider the effects of distributed load f_0. From (3.48),

$$\{f\} = \frac{f_0 L}{12} \begin{Bmatrix} 6 \\ -L \\ 6 \\ L \end{Bmatrix}. \tag{3.81}$$

The concentrated force and moment at the right end provide the following nodal forces at $x = L$, recalling from the sign conventions in Fig. 3.4 that shearing forces are positive when directed upward and bending moments are positive when clockwise in sense:

$$Q_3 = F_0, \qquad Q_4 = -M_0. \tag{3.82}$$

Now the contribution to nodal forces from the force F_1 applied at $x = L/3$ must be added. This contribution, labeled $\boldsymbol{F}_C$, can be extrapolated from the location of the point force to the nodal positions via use of the shape functions (3.30)–(3.33) as follows:

$$\{\boldsymbol{F}_C\} = F_1 \begin{Bmatrix} N_1(\xi) \\ N_2(\xi) \\ N_3(\xi) \\ N_4(\xi) \end{Bmatrix}_{\xi=1/3} = \frac{F_1}{27} \begin{Bmatrix} 20 \\ -4L \\ 7 \\ 2L \end{Bmatrix}. \tag{3.83}$$

Finally, superposition of forces in (3.81), (3.82), and (3.83) with unknown reactions at the wall gives the sought total force vector:

$$\{F\} = \{Q\} + \{f\} + \{F_C\} = \begin{Bmatrix} Q_1 \\ Q_2 \\ F_0 \\ -M_0 \end{Bmatrix} + \frac{f_0 L}{12} \begin{Bmatrix} 6 \\ -L \\ 6 \\ L \end{Bmatrix} + \frac{F_1}{27} \begin{Bmatrix} 20 \\ -4L \\ 7 \\ 2L \end{Bmatrix}. \tag{3.84}$$

Though not requested explicitly by the problem statement, the unknown deflection and rotation at the right end, u_3 and u_4 respectively, can be determined by solving the condensed equilibrium equations for a single beam element. The global (= single) element stiffness matrix prior to application of essential boundary conditions is that in (3.46), assuming constant EI. Since $u_1 = u_2 = 0$, the first two rows/columns can be eliminated, and the equilibrium equations to be solved, considering the corresponding bottom two entries of $\boldsymbol{F}$ in (3.84), are

$$\frac{2EI}{L^3} \begin{bmatrix} 6 & 3L \\ 3L & 2L^2 \end{bmatrix} \begin{Bmatrix} u_3 \\ u_4 \end{Bmatrix} = \begin{Bmatrix} F_0 + f_0 L/2 + 7F_1/27 \\ -M_0 + f_0 L^2/12 + 2F_1 L/27 \end{Bmatrix}. \tag{3.85}$$

3.7 Frame Elements

In this section, key features of frame elements are presented. A frame element can be interpreted as a modified beam element that supports axial loads in addition to transverse forces and applied bending moments. Individual frame elements can be arranged in various orientations in 2-D space and then combined to make a 2-D structure. The analog of a frame element in 3-D space would be a certain kind of plate or shell that supports in-plane and out-of-plane loads as well as bending moments about multiple axes. The study of plates and shells is outside the scope of this book.

3.7.1 Definitions

Recall from Chapter 2 that a truss element supports only axial loads. In contrast, a frame element potentially supports both axial and transverse loads, as well as concentrated bending moments. The distinction is illustrated by example in Fig. 3.10, which shows a two-element truss structure on the left and a two-element frame structure on the right. The intersection of the two truss elements is interpreted as a pin joint since no shearing forces or bending moments are allowed, while the intersection of the two frame elements is interpreted as a weld that potentially supports all three types of generalized forces: axial, transverse, and bending. As will be discussed shortly, a frame element can be constructed via superposition of features of a truss element and an Euler-Bernoulli beam element.

3.7.2 Matrix Equations

The superposition of a beam element with a 1-D truss element to produce a frame element is demonstrated in Fig. 3.11, which introduces notation for nodal degrees of freedom and nodal forces. In a local 2-D Cartesian coordinate system with the x- and y-axes aligned respectively parallel and perpendicular to the elements, a beam element includes four nodal degrees of freedom (deflection and rotation at each node), and a truss element includes two degrees of freedom (displacement at each node). The vector $\boldsymbol{u}$ of six generalized nodal displacements u_i for the frame is then defined as

$$\{\boldsymbol{u}\} = \begin{Bmatrix} u_1 \\ v_1 \\ \theta_1 \\ u_2 \\ v_2 \\ \theta_2 \end{Bmatrix}, \tag{3.86}$$

where subscripts 1 and 2 denote local node numbers. Analogously, the vector $\boldsymbol{Q}$ of six generalized nodal forces Q_i for the frame element is

$$\{\boldsymbol{Q}\} = \begin{Bmatrix} P_1 \\ V_1 \\ M_1 \\ P_2 \\ -V_2 \\ -M_2 \end{Bmatrix}. \tag{3.87}$$

Here, P_j, V_j, and M_j are respective axial forces, shearing forces, and bending moments, and j refers to the local node number. It is understood that all quantities are

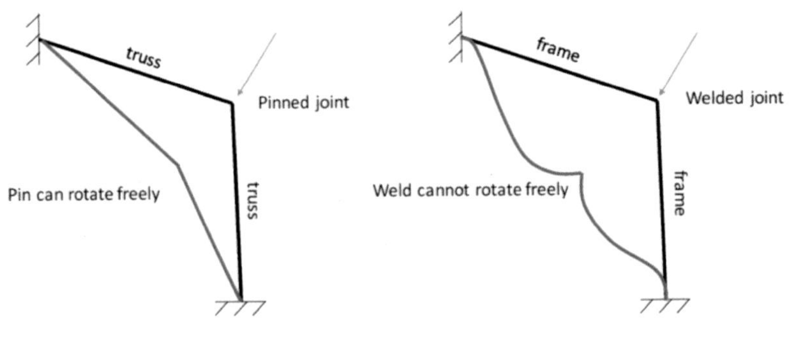

Fig. 3.10 2-D truss structure (left) compared to 2-D frame structure (right)

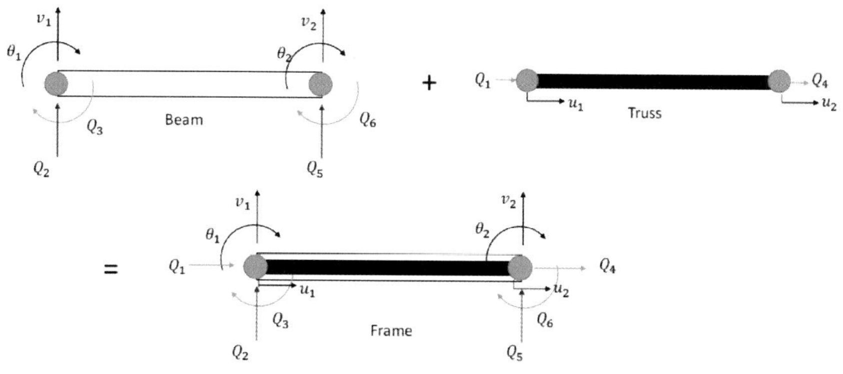

Fig. 3.11 Nodal degrees-of-freedom and generalized forces for beam, truss, and frame elements

referred to a single element; superscripts, e.g., e, are often suppressed on vectors and their components in this section when no possibility of confusion arises.

Attention is restricted to frame elements with constant elastic modulus E, constant cross-sectional area A, and constant moment of inertia I. The length of the element is denoted by L_e. In local coordinates, recall from (2.60) that the stiffness matrix of a 2-D truss element is

$$[\boldsymbol{k}^{(e)}_{\text{truss}}] = \frac{EA}{L_e}\begin{bmatrix} 1 & 0 & -1 & 0 \\ 0 & 0 & 0 & 0 \\ -1 & 0 & 1 & 0 \\ 0 & 0 & 0 & 0 \end{bmatrix}. \tag{3.88}$$

Recall also from (3.46) that in local coordinates, the stiffness matrix of an Euler-Bernoulli beam element is

$$[\boldsymbol{k}^{(e)}_{\text{beam}}] = \frac{2EI}{L_e^3}\begin{bmatrix} 6 & -3L_e & -6 & -3L_e \\ -3L_e & 2L_e^2 & 3L_e & L_e^2 \\ -6 & 3L_e & 6 & 3L_e \\ -3L_e & L_e^2 & 3L_e & 2L_e^2 \end{bmatrix}. \tag{3.89}$$

Taking into account the definitions of generalized displacements and their conjugate forces in (3.86) and (3.87), a 6×6 generalized stiffness matrix that combines (3.88) and (3.89) is constructed as follows in the local coordinate system (c.s.) aligned with the frame element:

$$[\boldsymbol{k}^{(e)}_{\text{frame}}] = \frac{2EI}{L_e^3}\begin{bmatrix} \mu & 0 & 0 & -\mu & 0 & 0 \\ 0 & 6 & -3L_e & 0 & -6 & -3L_e \\ 0 & -3L_e & 2L_e^2 & 0 & 3L_e & L_e^2 \\ -\mu & 0 & 0 & \mu & 0 & 0 \\ 0 & -6 & 3L_e & 0 & 6 & 3L_e \\ 0 & -3L_e & L_e^2 & 0 & 3L_e & 2L_e^2 \end{bmatrix}, \qquad \mu = \frac{AL_e^2}{2I}, \quad \text{(local c.s.)} \tag{3.90}$$

Restricting attention to distributed axial load q and transverse load f that are constant over the element, and denoting by $\boldsymbol{F}$ the total load vector consisting of concentrated loads $\boldsymbol{Q}$ and contributions of these uniform distributed loads $\boldsymbol{f}$, inserting the stiffness matrix of (3.90) into the element equilibrium equations for a single linear frame element produces

$$[\boldsymbol{k}^{(e)}_{\text{frame}}]\{\boldsymbol{u}\} = \{\boldsymbol{Q}\} + \{\boldsymbol{f}\} = \{\boldsymbol{F}\} \rightarrow$$

$$\frac{2EI}{L_e^3}\begin{bmatrix} \mu & 0 & 0 & -\mu & 0 & 0 \\ 0 & 6 & -3L_e & 0 & -6 & -3L_e \\ 0 & -3L_e & 2L_e^2 & 0 & 3L_e & L_e^2 \\ -\mu & 0 & 0 & \mu & 0 & 0 \\ 0 & -6 & 3L_e & 0 & 6 & 3L_e \\ 0 & -3L_e & L_e^2 & 0 & 3L_e & 2L_e^2 \end{bmatrix} \begin{Bmatrix} u_1 \\ u_2 \\ u_3 \\ u_4 \\ u_5 \\ u_6 \end{Bmatrix} = \begin{Bmatrix} Q_1 \\ Q_2 \\ Q_3 \\ Q_4 \\ Q_5 \\ Q_6 \end{Bmatrix} + \begin{Bmatrix} \frac{1}{2}qL_e \\ \frac{1}{2}fL_e \\ -\frac{1}{12}fL_e^2 \\ \frac{1}{2}qL_e \\ -\frac{1}{2}fL_e \\ \frac{1}{12}fL_e^2 \end{Bmatrix} = \begin{Bmatrix} F_1 \\ F_2 \\ F_3 \\ F_4 \\ F_5 \\ F_6 \end{Bmatrix}. \tag{3.91}$$

Equations (2.128) and (3.48) have been consulted in derivation of the vector $\boldsymbol{f}$ that combines nodal effects of constant distributed loads applied on linear truss and beam elements, respectively.

3.7.3 Coordinate Transformations

When considering a frame structure comprised of multiple elements that are not all oriented parallel in the ambient space, the local element quantities must be referred to a single global coordinate system prior to assembly. Notation associated with a frame element in local coordinates aligned with its axial and transverse directions is shown on the left side of Fig. 3.12. Notation in global xy-coordinates is shown on the right. The angle α transforms the global coordinate frame to the local frame when defined as positive in a counterclockwise sense. The distributed transverse and axial forces, both measured per unit length, are written as f and q, respectively, and are always referred to local coordinates. In local coordinates, nodal degrees of freedom and nodal forces are written with overbars, i.e., column vectors $\bar{\boldsymbol{u}}$ and $\bar{\boldsymbol{Q}}$. In global coordinates, the analogous variables are written simply as $\boldsymbol{u}$ and $\boldsymbol{Q}$. Note also that rotations are the same in both systems, i.e., $\theta = \bar{\theta}$, positive in sign when oriented clockwise in sense.

Transformation of vectors from global to local coordinates for 2-D frame elements follows a procedure analogous to that for 2-D truss elements of §2.2.8.

Specifically, nodal displacements and nodal forces transform from global to local coordinates as

$$\{\bar{\boldsymbol{u}}\} = [\boldsymbol{T}^{(e)}]\{\boldsymbol{u}\}, \qquad \{\bar{\boldsymbol{Q}}\} = [\boldsymbol{T}^{(e)}]\{\boldsymbol{Q}\}, \tag{3.92}$$

where $\boldsymbol{T}^{(e)}$ is a 6×6 orthogonal transformation matrix. Since $[\boldsymbol{T}^{(e)}]^{-1} = [\boldsymbol{T}^{(e)}]^{\mathrm{T}}$, degrees of freedom and force vectors change from local to global coordinates as follows:

$$\{\boldsymbol{u}\} = [\boldsymbol{T}^{(e)}]^{\mathrm{T}}\{\bar{\boldsymbol{u}}\}, \qquad \{\boldsymbol{Q}\} = [\boldsymbol{T}^{(e)}]^{\mathrm{T}}\{\bar{\boldsymbol{Q}}\}. \tag{3.93}$$

The transformation matrix for frame elements is

$$[\boldsymbol{T}^{(e)}] = \begin{bmatrix} \cos\alpha & \sin\alpha & 0 & 0 & 0 & 0 \\ -\sin\alpha & \cos\alpha & 0 & 0 & 0 & 0 \\ 0 & 0 & 1 & 0 & 0 & 0 \\ 0 & 0 & 0 & \cos\alpha & \sin\alpha & 0 \\ 0 & 0 & 0 & -\sin\alpha & \cos\alpha & 0 \\ 0 & 0 & 0 & 0 & 0 & 1 \end{bmatrix}. \tag{3.94}$$

Following the same steps as used in §2.2.8, a relationship between the element stiffness matrix in local coordinates, $\bar{\boldsymbol{k}}^{(e)}_{\text{frame}}$, and that in global coordinates, $\boldsymbol{k}^{(e)}_{\text{frame}}$, is derived:

$$[\boldsymbol{k}^{(e)}_{\text{frame}}] = [\boldsymbol{T}^{(e)}]^{\mathrm{T}}[\bar{\boldsymbol{k}}^{(e)}_{\text{frame}}][\boldsymbol{T}^{(e)}] \Leftrightarrow [\bar{\boldsymbol{k}}^{(e)}_{\text{frame}}] = [\boldsymbol{T}^{(e)}][\boldsymbol{k}^{(e)}_{\text{frame}}][\boldsymbol{T}^{(e)}]^{\mathrm{T}}. \tag{3.95}$$

Using (3.90) for the stiffness in the local element coordinate system, invoking the shorthand notation $c = \cos\alpha$ and $s = \sin\alpha$, and substituting (3.94) into (3.95), the element stiffness matrix in global coordinates is computed as

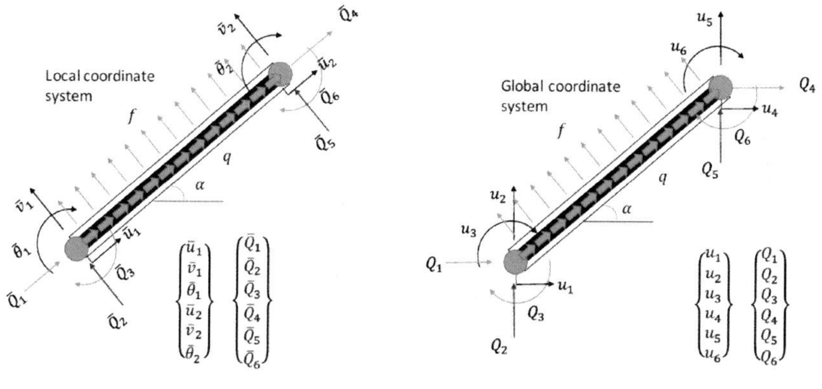

Fig. 3.12 Frame element quantities referred to local (left) and global (right) coordinate systems

$$[\boldsymbol{k}^{(e)}_{\text{frame}}] = \frac{2EI}{L_e^3}\begin{bmatrix} \mu c^2+6s^2 & (\mu-6)cs & 3L_e s & -\mu c^2-6s^2 & (\mu+6)cs & 3L_e s \\ (\mu-6)cs & \mu s^2+6c^2 & -3L_e c & (\mu+6)cs & -\mu s^2-6c^2 & -3L_e c \\ 3L_e s & -3L_e c & 2L_e^2 & 3L_e s & 3L_e c & L_e^2 \\ -\mu c^2-6s^2 & (6-\mu)cs & 3L_e s & \mu c^2+6s^2 & (\mu-6)cs & -3L_e s \\ (6-\mu)cs & -\mu s^2-6c^2 & 3L_e c & (\mu-6)cs & \mu s^2+6c^2 & 3L_e c \\ 3L_e s & -3L_e c & L_e^2 & -3L_e s & 3L_e c & 2L_e^2 \end{bmatrix}. \tag{3.96}$$

The remainder of the analysis of a frame problem using FEA follows the same overall procedure as that for 2-D truss elements. Specifically, the remaining steps are

- **Assembly**: using the element connectivity information, combine the stiffness matrices, displacement vectors, and force vectors of each element into a global system of equations, with all quantities referred to the single global coordinate system;
- **Boundary Conditions**: manipulate the global equations to account for essential boundary conditions on primary solution variables, moving contributions of nonzero imposed generalized displacements to the right side of the global system of equations;
- **Solve**: solve the condensed global system $[\boldsymbol{K}]\{\boldsymbol{u}\} = \{\boldsymbol{F}\}$ where boundary conditions have been imposed, such that condensed stiffness matrix $[\boldsymbol{K}]$ is non-singular and therefore invertible;
- **Post-process**: compute secondary nodal and elemental quantities by substituting primary solution variables ($\boldsymbol{u}$) into element equations, where vectors and matrices are referred to a consistent coordinate system (i.e., a local or the global system as appropriate).

Since the response of the material is modeled as linear elastic, stresses in frame elements can be calculated by superposing results from the truss and beam equations. For example, total axial stresses result from axial extension/compression (i.e., truss-type contributions) and bending (i.e., beam-type contributions).

3.8 ANSYS Example: Cantilevered Beam

The forthcoming example demonstrates concepts from beam theory using the ANSYS software (v.17.2, 2016). The problem is intentionally simple to enable validation of its numerical solution with the exact result from Euler-Bernoulli beam theory.

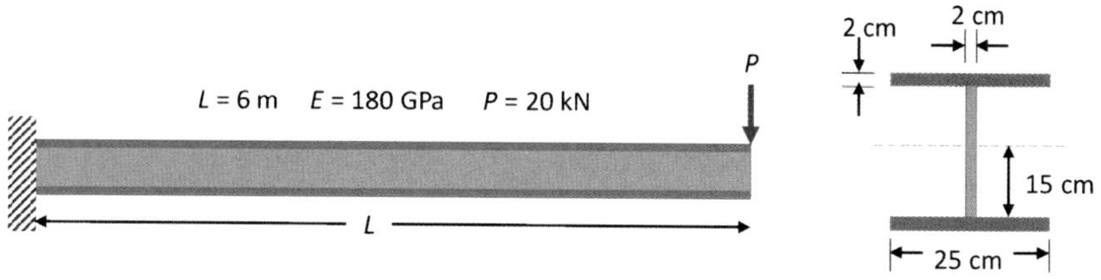

Fig. 3.13 Cantilevered beam example problem: geometry, boundary conditions, and properties

3.8.1 Problem Statement

Consider the cantilevered beam shown in Fig. 3.13. The beam is fully constrained at its left end at $x = 0$, with corresponding null displacement and null rotation. A point force of magnitude $P = 20$ kN is applied transversely downward at the right end at $x = L = 6$ m. This end is free to rotate. The elastic modulus of the beam is $E = 180$ GPa, and the geometry and dimensions of the cross section are shown on the right side of the figure. The main objectives are determination of the maximum deflection, maximum rotation, and maximum tensile stress supported by the beam. A secondary objective is validation of the results obtained from the FE software.

3.8.2 Pre-processing

Pre-processing encompasses selection of element type and material properties, prescription of beam length and cross-sectional geometric properties, and mesh creation. The FE analysis in the ANSYS package is initiated by selection of the element type suitable for a long/slender beam, the `beam 188` element, by the following sequence:

```
Main Menu > Preprocessor > Element Type > Add/Edit/Delete
> Add > Beam > beam 188
```

See Fig. 3.14 which also includes options specified for the element type. This element allows consideration of nonlinearities associated with large deformation and is somewhat more sophisticated[1] than the Euler-Bernoulli elements featured in the previous main text of Chapter 3, but it is still sufficient for solving linear problems. Next, the material model is prescribed as isotropic and linear elastic:

[1] Specifically, the `beam 188` element invokes Timoshenko's beam theory whose detailed description is beyond the scope of this book.

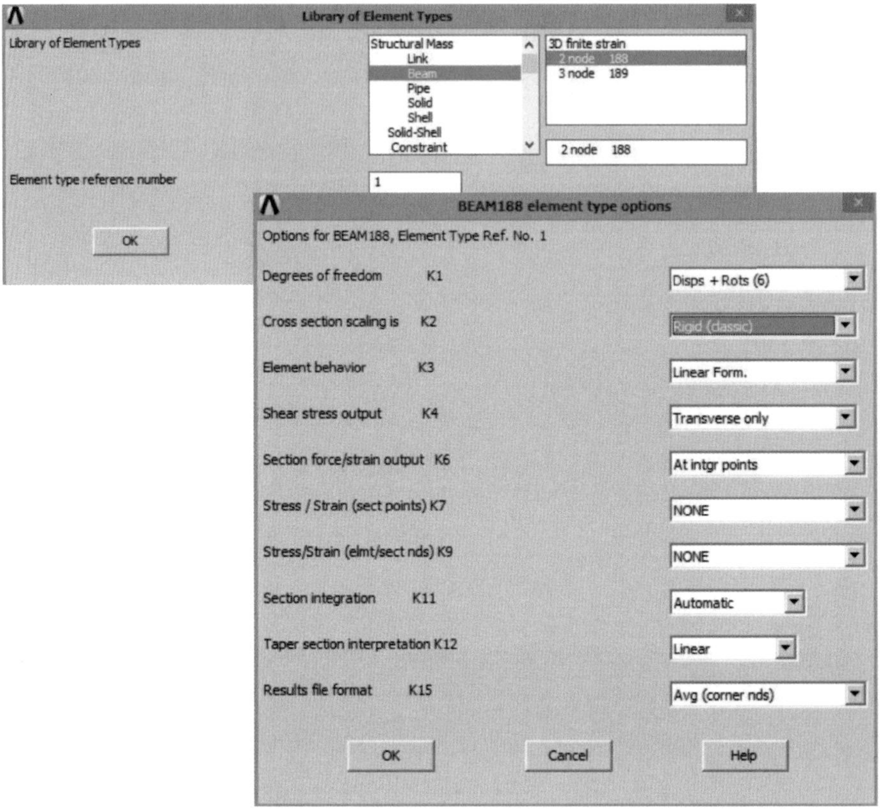

Fig. 3.14 Cantilevered beam example problem: element data

```
Main Menu > Preprocessor > Material Properties > Material
Models > Structural > Linear > Elastic > Isotropic
```

The software requests an elastic modulus E and Poisson's ratio ν. The former is prescribed as 180e9 Pa, while the latter is arbitrary for this problem and can be safely chosen as 0.25, for example, though any value obeying $-1 < \nu < 0.5$ is acceptable. We use SI units for consistency. The material definition procedure is shown in Fig. 3.15. A section must be created to specify geometric properties of the cross section:

```
Main Menu > Preprocessor > Sections > Beam > Common
Sections
```

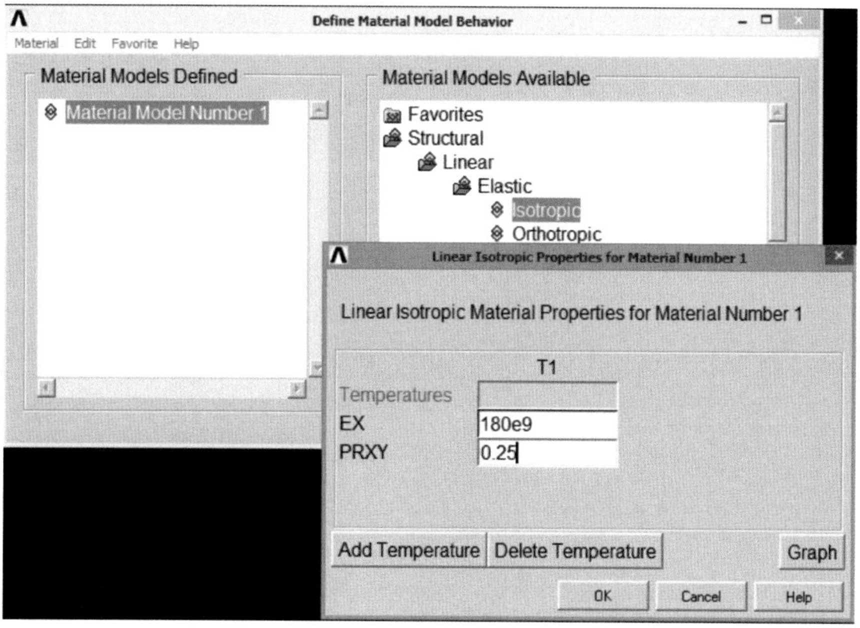

Fig. 3.15 Cantilevered beam example problem: material data

Enter the appropriate cross-sectional parameters from Fig. 3.13 as shown on the left side of Fig. 3.16. The ID and name of the section can be entered arbitrarily; since all elements have the same properties, each element will be assigned to this section by default. The geometry along the length of the beam in the x-direction is specified as follows. In ANSYS, the default coordinate system for beam elements oriented along the x-direction aligns the transverse deflection in the z-direction rather than the y-direction used earlier in the theory sections of this chapter. Here, we first create two keypoints located at (0,0,0) and (6,0,0), recalling that $L = 6$ m:

```
Main Menu > Preprocessor > Modeling > Create > Keypoints >
In Active CS
```

Then we create a line between these two points:

```
Main Menu > Preprocessor > Modeling > Create > Lines >
Lines > Straight Line
```

Pick the two keypoints and click "OK". A mesh of 10 elements is used. To generate such a mesh, first set the mesh size as $L/10 = 0.6$ m:

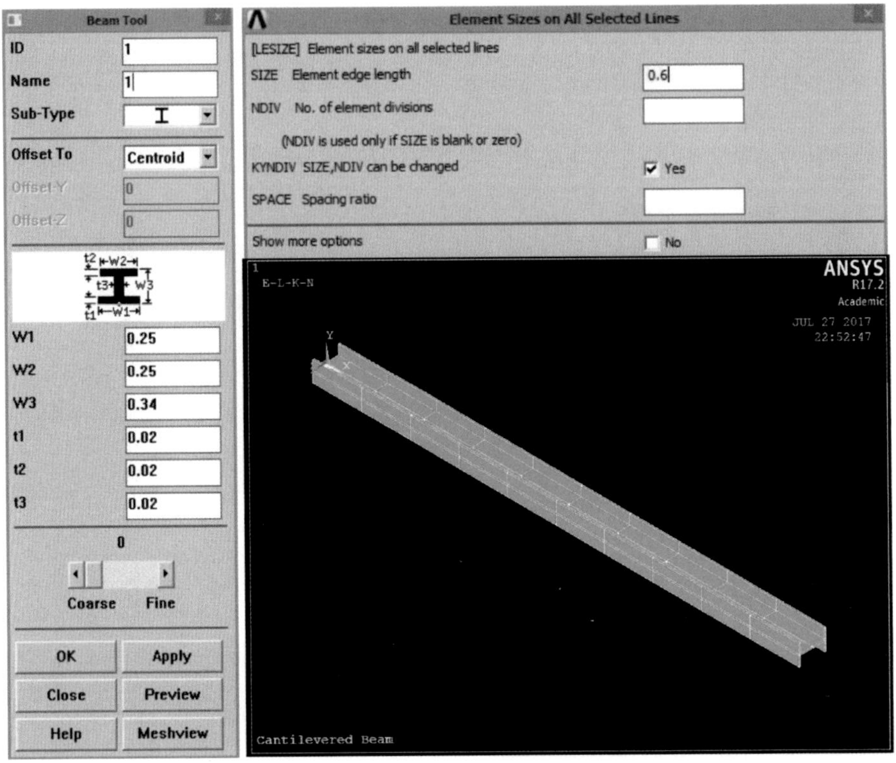

Fig. 3.16 Cantilevered beam example problem: section and elements

```
Main Menu > Preprocessor > Meshing > Size Cntrls > Manual
Size > Lines > All Lines
```

Then mesh the line via

```
Main Menu > Preprocessor > Mesh > Lines > Pick
```

The mesh of 10 elements can be visualized with the cross-sectional geometry by invoking the following sequence:

```
Main Menu > PlotCtrls > Style > Size and Shape > ESHAPE
```

Check "On". The end result is shown on the right side of Fig. 3.16 in an isometric view. It is understood that fewer elements would be sufficient to accurately solve

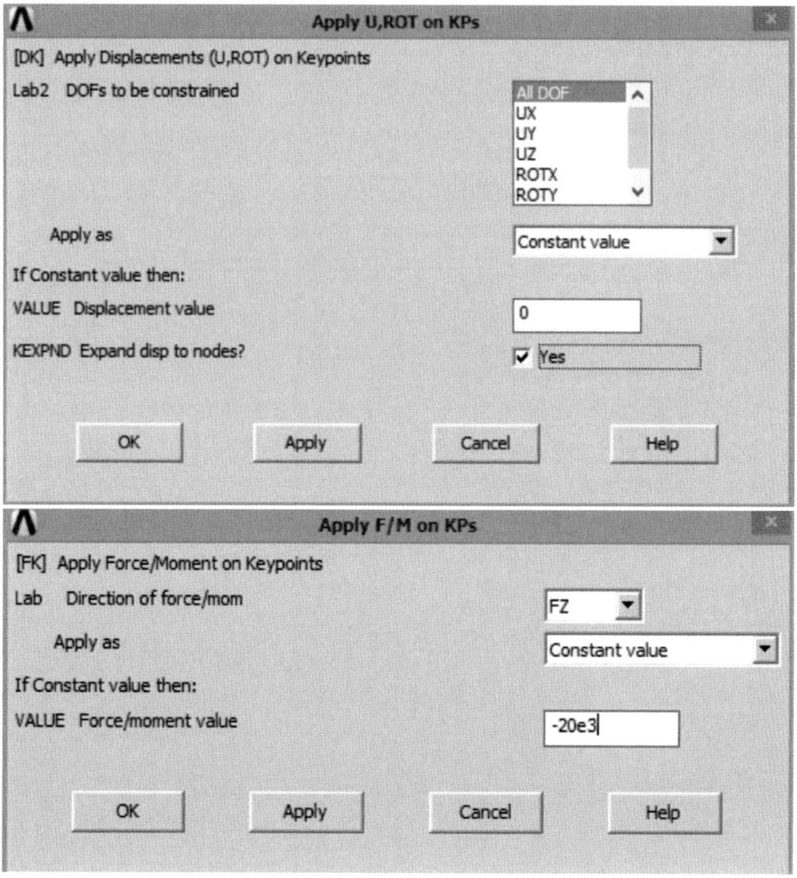

Fig. 3.17 Cantilevered beam example problem: boundary conditions

this problem since the exact solution for transverse deflection is a cubic function of x, as will be shown later, and since the element shape functions for a beam element with two nodes are cubic as explained in §3.3. Regardless, multiple elements are used here to demonstrate the meshing procedure in ANSYS.

3.8.3 Solution

The steps for this solving this example problem are the following: define the problem type, assign boundary conditions, and select possible solver options. The analysis type is static, specified in ANSYS as

```
Main Menu > Solution > Analysis Type > New Analysis >
Static
```

Boundary conditions are invoked here in the solution module, but they could equivalently be prescribed in the pre-processing module. First, displacement conditions, i.e., essential BCs, are enforced at keypoint 1 at the left end:

```
Main Menu > Solution > Define Loads > Apply > Structural >
Displacement > On Keypoints
```

Constrain all degrees of freedom (DOF) to be zero at keypoint 1, as shown in Fig. 3.17. The natural BC corresponding to the point force P at $x = L$ is then prescribed:

```
Main Menu > Solution > Define Loads > Apply > Structural >
Force/Moment > On Keypoints
```

Enter the FZ value of $P = -20$ kN as shown in Fig. 3.17. The default solver options are adequate for this problem. These are displayed by the pre-check that is executed via the sequence below:

```
Main Menu > Solution > Solve > Current LS > OK
```

Click "OK" to execute the solver and obtain the FE solution.

3.8.4 Post-processing

Recalling the coordinate system used in ANSYS, the primary solution to be investigated is the transverse displacement υ and the rotation about the y-axis, θ, for each of the eleven nodes. The following GUI selections create a list of displacement and rotation for every node:

```
Main Menu > General Postproc > List Results > Nodal
Solution > DOF Solution > Z-Component of displacement
```

```
Main Menu > General Postproc > List Results > Nodal
Solution > DOF Solution > Y-Component of rotation
```

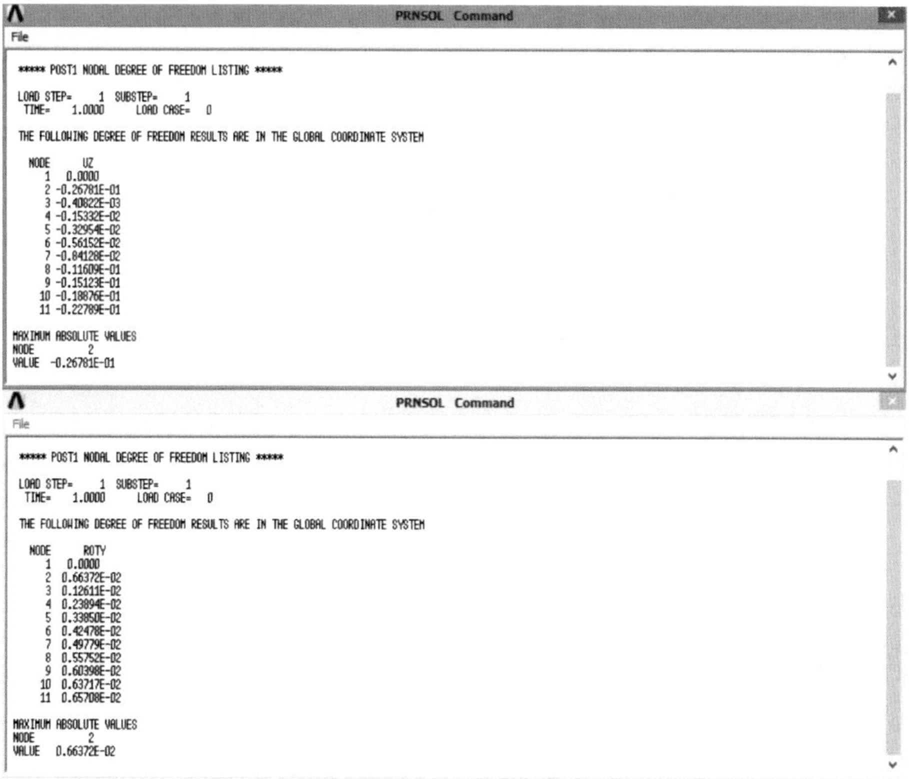

Fig. 3.18 Cantilevered beam example problem: primary solution

See Fig. 3.18 for the listings of results. The deformed shape of the beam can be viewed in superposition with the original unloaded beam:

```
Main Menu > General Postproc > Plot Results > Deformed
Shape
```

The displacements are magnified by default in the ANSYS GUI to enable visualization; otherwise, the original and deformed geometries would be indistinguishable. A contour plot of the axial/bending stress in the beam is generated by

```
Main Menu > General Postproc > Plot Results > Element
Solution > X-Component of stress
```

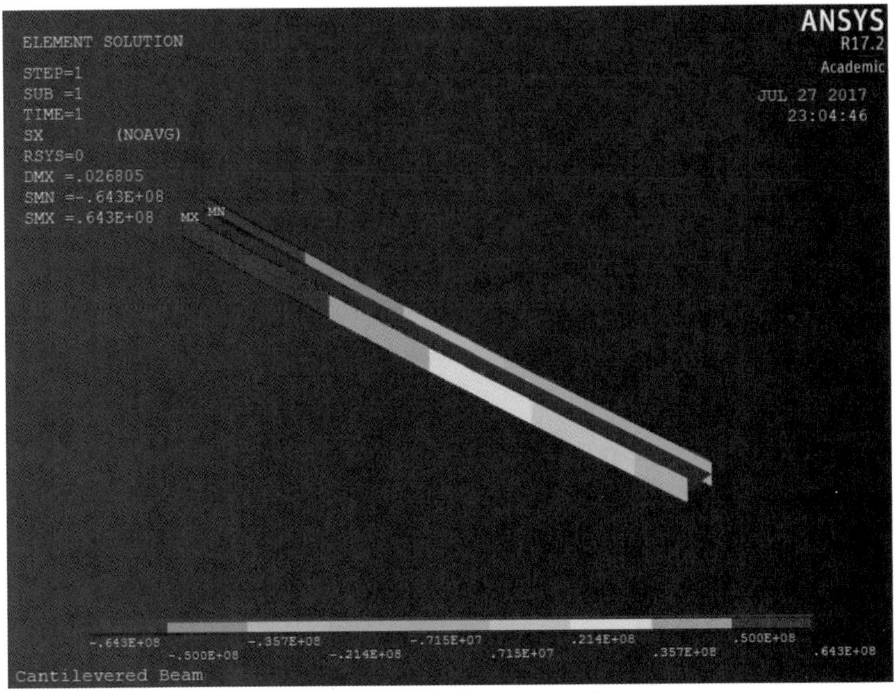

Fig. 3.19 Cantilevered beam example problem: stress contours

The stress is tensile or positive along the left edge and compressive or negative along the right edge, as expected from physical intuition noting that the applied point force is directed from left to right.

The numerical results from this problem can be validated versus the exact solution of the strong form differential equation of Euler-Bernoulli beam theory, (3.15), and supplementary equations of §3.1 and §3.2. The strong form equation becomes the following fourth-order ODE with constant E and constant I, in the absence of distributed loads:

$$EI\frac{\mathrm{d}^4 \upsilon}{\mathrm{d}x^4} = 0. \tag{3.97}$$

Integrating four times provides the polynomial solution

$$EI\upsilon(x) = C_0 + C_1 x + C_2 x^2 + C_3 x^3. \tag{3.98}$$

The coefficients C_i, $i = 0, \ldots, 3$ are determined from the essential boundary conditions at the left end,

$$\upsilon(0) = 0, \qquad \theta(0) = -\frac{\mathrm{d}\upsilon}{\mathrm{d}x}(0) = 0, \tag{3.99}$$

giving $C_0 = C_1 = 0$, and the natural boundary conditions at the right end,

$$M(L) = EI\frac{\mathrm{d}^2 \upsilon}{\mathrm{d}x^2}(L) = 0, \qquad V(L) = EI\frac{\mathrm{d}^3 \upsilon}{\mathrm{d}x^3}(L) = P, \tag{3.100}$$

giving $C_2 = -PL/2$ and $C_3 = P/6$. The resulting exact analytical solution is thus

$$\upsilon(x) = -\frac{Px^2}{6EI}(3L - x), \qquad \theta(x) = \frac{Px}{2EI}(2L - x); \tag{3.101}$$

$$M(x) = -P(L - x), \qquad V = P. \tag{3.102}$$

The bending stress is, from (3.62),

$$\sigma(x) = -\frac{M(x)y}{I} = \frac{P(L - x)y}{I}. \tag{3.103}$$

For this particular problem, the moment of inertia for the cross section of the beam (I-beam) about the neutral axis is computed as [1]

$$\begin{aligned} I &= 2\left[\frac{1}{12}(0.25)(0.02)^3 + (0.25)(0.02)(0.16)^2\right] + \frac{1}{12}(0.02)(0.3)^3 \text{ m}^4 \\ &= 3.013 \times 10^{-4} \text{ m}^4. \end{aligned} \tag{3.104}$$

The maximum magnitudes of deflection and rotation occur at the right end, where $x = L$:

$$\upsilon(L) = -\frac{PL^3}{3EI} = -2.66 \text{ cm}, \qquad \theta(L) = \frac{PL^2}{2EI} = 6.64 \times 10^{-3} \text{ rad}. \tag{3.105}$$

The most severe bending moment and maximum tensile stress arise at the left end, at $x = 0$, and along the upper edge ($y = 0.17$ m):

$$M(0) = -PL = -12 \text{ kN}\cdot\text{m}, \quad \sigma(x = 0, y = 0.17 \text{ m}) = \frac{PLy}{I} = 67.7 \text{ MPa}. \tag{3.106}$$

From Fig. 3.18, the maximum (negative) deflection obtained from ANSYS is -2.68 cm at node 2 corresponding to $x = L$, and the maximum magnitude of rotation is 6.64×10^{-3} rad at node 2. From Fig. 3.19, the maximum bending stress is 64.3 MPa at node 1 corresponding to $x = 0$. These values are very close, or equal for the case of θ, to the analytical solution in (3.105) and (3.106), thus validating the FE procedure and numerical results. Slight differences may arise as a result of the different (Timoshenko) beam theory invoked in the software that considers some shear deformation and rotational bending effects absent in Euler-Bernoulli beam theory.

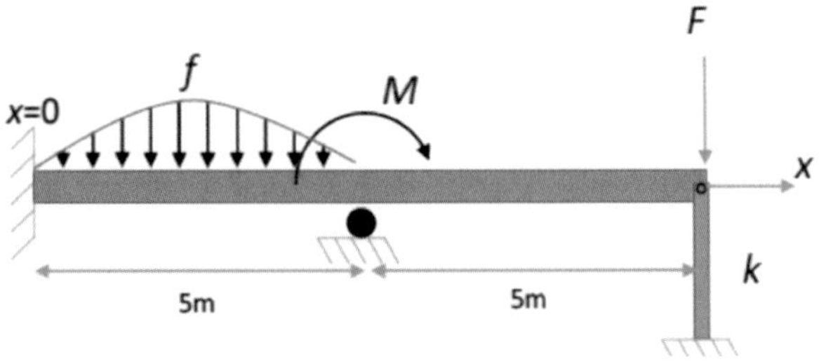

Fig. 3.20 Problem 3.9.1

3.9 Problems

3.9.1. Consider the truss and Euler-Bernoulli beam in the FE problem depicted in Fig. 3.20. The truss has a constant stiffness of k, and the beam has a constant stiffness (modulus times moment of inertia) of EI. A concentrated force F, a concentrated moment M, and a distributed load f are applied. The distributed load is of the following mathematical form, with f_0 a constant:

$$f(x) = f_0 \left[1 - \frac{(x-2.5)^2}{(2.5)^2} \right].$$

a. Determine the minimum number of finite elements to use and then find the assembled stiffness matrix.
b. Determine the element force vectors and the assembled force vector.
c. Apply boundary conditions and determine the condensed system of vector-matrix equilibrium equations.
d. Determine the displacement and rotation at each node using the following numerical values: $F = 4$ MN, $M = 0.7$ MN·m, $f_0 = 3$ MN/m, $k = 0.7$ MN/m, and $EI = 1$ GN·m^2.
e. Compute the shearing forces and bending moments at each node of each element using geometric and material parameters from part **d**.
f. Evaluate the bending moment at $x = 7.5$ m using the same set of parameters from part **d**.

3.9.2. This problem involves a cantilevered beam 1 m long with a square cross-section of dimensions 10 cm $\times$10 cm. Young's modulus $E = 100$ GPa. A distributed load of magnitude 10 kN/m is applied downward along the entire length of the beam. [A "cantilevered" beam is fully clamped at one end.]

a. Perform calculations in each part i, ii, and iii below using a finite element software package. Use the mesh characteristics below and determine the deflection and rotation at the free end for each case. List the results in a table. This table does not have to be created directly using the FEM software, but simply contains the numerical results obtained from the software. Explain any differences you observe among the different cases, consulting the software help manual as needed.

i. A beam mesh comprised of 1 element. First use equivalent nodal loads to represent the distributed load. Then use the beam distributed pressure load.
ii. A beam mesh comprised of 10 equal sized elements. Use the beam distributed pressure load.
iii. A solid 3-D mesh comprised of 10 equal sized hexahedral (brick) elements. Use a distributed pressure load and a Poisson's ratio of $\nu = 0.25$.

b. Evaluate the exact solution using Euler-Bernoulli (EB) beam theory. [Hint: integrate the 4th order ODE from the strong form and use BCs to find integration constants]. Compare and thereby validate each of the numerical solutions from part **a**. with the exact EB beam theory results. Report the differences you observe and comment on possible reasons for differences.

References

1. R.C. Hibbeler, *Mechanics of Materials* (10th edition, Pearson, Boston, 2016)
2. J.N. Reddy, *An Introduction to the Finite Element Method* (2nd edition, McGraw-Hill, New York, 1993)

Chapter 4
Planar Two-Dimensional Analysis

Abstract Two-dimensional (2-D) problems are considered in this chapter, wherein generally, field variables depend on the two Cartesian spatial coordinates x and y. Described first are heat conduction problems wherein the primary field variable is the scalar temperature field. Governing equations and boundary conditions are presented, followed by derivation of the weak form equations. Interpolation functions for 2-D analyses are introduced. Matrix equations of thermal equilibrium are derived, with discretized forms and flux boundary conditions then addressed in detail. Assembly, solution, and post-processing are discussed. Then, planar elasticity problems are described in the context of FEA, including plane stress and plane strain conditions, with the 2-D displacement vector field depending generally on both x and y. Governing equations, strong and weak forms, and matrix equations are then derived and discussed.

Heat transfer problems are addressed first in this chapter. The second half of the chapter addresses solid mechanics problems. In both cases, attention is restricted to planar 2-D problems in static equilibrium or at a steady state, as opposed to dynamic or transient conditions that will be considered in a later chapter.

4.1 Heat Conduction: Governing Equations

Consider a 2-D domain Ω with boundary $\partial\Omega$, where Cartesian coordinates x and y cover the domain. As shown in Fig. 4.1, the unit outward normal vector to Ω is $\boldsymbol{n}(x,y)$. Denote the temperature field by $T(x,y)$, the heat flux vector by $\boldsymbol{q}(x,y)$, and the scalar heat source by $f(x,y)$.

Boundary conditions on $\partial\Omega$ are of two possible types:

- **Essential**: temperature prescribed on $\partial\Omega_T$;
- **Natural**: scalar heat flux prescribed on $\partial\Omega_q$.

The scalar heat flux on the boundary is defined as $q_n = \boldsymbol{q}\cdot\boldsymbol{n}$.

The local balance of energy for static heat conduction, i.e., thermal equilibrium, is written in vector form as [1, 3]

$$\nabla \cdot \boldsymbol{q} = f, \tag{4.1}$$

where $\nabla(\cdot)$ is the gradient operator and "$\cdot$" the scalar product of vectors, such that $\nabla \cdot (\cdot)$ is the divergence of a vector field. In 2-D Cartesian coordinates, this equation is

$$\frac{\partial q_x(x,y)}{\partial x} + \frac{\partial q_y(x,y)}{\partial y} = f(x,y). \tag{4.2}$$

Heat flux $\boldsymbol{q}$ by convention used in this book is positive in sign in the direction of energy flow, such that q_n is positive when thermal energy leaves the body. Scalar heat source f is positive in sign if it would tend to increase the internal energy of the body.

The constitutive model considered herein is linear Fourier heat conduction. Denote by $\boldsymbol{k}(x,y)$ the symmetric thermal conductivity matrix, of size 2×2 for planar problems:

$$[\boldsymbol{k}] = \begin{bmatrix} k_{11} & k_{12} \\ k_{12} & k_{22} \end{bmatrix} = \begin{bmatrix} k_{xx} & k_{xy} \\ k_{xy} & k_{yy} \end{bmatrix}. \tag{4.3}$$

When $\boldsymbol{k} = k\mathbf{I}$ (i.e., $k_{ij} = k\delta_{ij}$) with $k(x,y)$ a scalar, the heat conduction material properties become isotropic, i.e., directionally invariant. Fourier's law is the following constitutive equation in vector-matrix form:

$$\{\boldsymbol{q}\} = -[\boldsymbol{k}]\{\nabla T\} \leftrightarrow \begin{Bmatrix} q_x \\ q_y \end{Bmatrix} = -\begin{bmatrix} k_{xx} & k_{xy} \\ k_{xy} & k_{yy} \end{bmatrix} \begin{Bmatrix} \partial T/\partial x \\ \partial T/\partial y \end{Bmatrix}. \tag{4.4}$$

With components of $\boldsymbol{k}$ generally positive in sign, the positive heat flux is typically directed from hot to cold locations of the body. Dimensions of heat flux, heat source, and thermal conductivity are respectively energy rate per unit area, energy rate per unit volume, and energy rate per unit length per unit temperature.

The strong form of the problem statement for steady state conduction is obtained by substituting (4.4) into (4.1):

$$-\nabla \cdot (\boldsymbol{k}\nabla T) = f, \tag{4.5}$$

or in 2-D scalar form:

$$-\frac{\partial}{\partial x}\left(k_{11}\frac{\partial T}{\partial x} + k_{12}\frac{\partial T}{\partial y}\right) - \frac{\partial}{\partial y}\left(k_{12}\frac{\partial T}{\partial x} + k_{22}\frac{\partial T}{\partial y}\right) = f. \tag{4.6}$$

When the conductivity is both constant and isotropic, i.e., when $\boldsymbol{k} = k\mathbf{I}$ with k =constant, (4.6) reduces to the Poisson equation:

$$\frac{\partial^2 T}{\partial x^2} + \frac{\partial^2 T}{\partial y^2} = -\frac{f}{k} \leftrightarrow \nabla^2 T = -f/k. \tag{4.7}$$

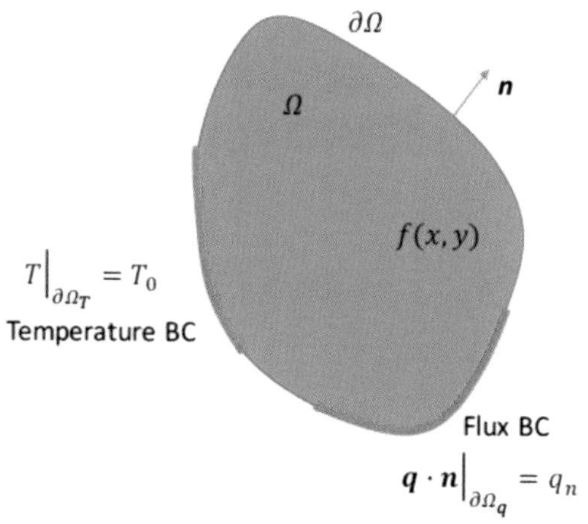

Fig. 4.1 Domain Ω for heat conduction analysis in two dimensions with boundary conditions

The Laplacian operator is $\nabla^2(\cdot)$. Furthermore, (4.7) degenerates to the Laplace equation when scalar point sources vanish:

$$\partial^2 T/\partial x^2 + \partial^2 T/\partial y^2 = 0 \leftrightarrow \nabla^2 T = 0. \tag{4.8}$$

In this particular simplest case, the steady state solution does not depend on the value of k.

Most of the treatment in this chapter will address the more general (anisotropic, nonzero f) case in (4.6). The complete strong form for the problem statement corresponding to Fig. 4.1 seeks the temperature field $T(x,y)$ in domain Ω that satisfies second-order partial differential equation (4.6), given boundary conditions on $\partial\Omega$, imposed heat source field f, and conductivity matrix $\boldsymbol{k}$ that may generally depend on position.

4.2 Heat Conduction: Weak Form

Derivation of the weak form statement for 2-D heat transfer problems follows a similar procedure as that for geometrically 1-D domains for trusses/bars and beams in Chapters 2 and 3, respectively. The present derivation is a bit more mathematically cumbersome than the former cases, however, since integration by parts necessarily involves consideration of area and contour integrals and use of the generalized theorem of Gauss.

Prior to derivation of the particular weak form equation for heat conduction, we review several identities from vector calculus that will be needed later. The first is the generalized theorem of Gauss, alternatively referred to in various literature as Green's theorem or a form of the divergence theorem. Consider a compact domain Ω with unit outward normal $\boldsymbol{n}$ on boundary $\partial\Omega$, and let $\boldsymbol{v}$ be a continuously differentiable vector field in this domain. Then the generalized theorem of Gauss is, in coordinate-free notation [1, 2]

$$\int_{\Omega} \nabla \cdot \boldsymbol{v} \, \mathrm{d}\Omega = \oint_{\partial\Omega} \boldsymbol{v} \cdot \boldsymbol{n} \, \mathrm{d}\partial\Omega, \tag{4.9}$$

where $\oint(\cdot)$ denotes integration over the boundary enclosing Ω. In 2-D, let area Ω be covered by Cartesian coordinates (x, y), and let s denote arc length along the curve $\partial\Omega$ enclosing Ω. Then in the present 2-D context, (4.9) becomes

$$\int_{\Omega} \left[\frac{\partial v_x(x,y)}{\partial x} + \frac{\partial v_y(x,y)}{\partial y} \right] \mathrm{d}x\mathrm{d}y = \oint_{\partial\Omega} (v_x n_x + v_y n_y) \mathrm{d}s. \tag{4.10}$$

Now we introduce a generic differentiable scalar field $a(x, y)$. A version of integration by parts can then be derived via use of (4.9) and applying the product rule for the gradient operator on $a\boldsymbol{v}$:

$$\int_{\Omega} a \nabla \cdot \boldsymbol{v} \, \mathrm{d}\Omega = \oint_{\partial\Omega} a \boldsymbol{v} \cdot \boldsymbol{n} \, \mathrm{d}\partial\Omega - \int_{\Omega} \nabla a \cdot \boldsymbol{v} \, \mathrm{d}\Omega. \tag{4.11}$$

In 2-D component form, the integration by parts identity in (4.11) reduces to the following equality:

$$\int_{\Omega} a \left(\frac{\partial v_x}{\partial x} + \frac{\partial v_y}{\partial y} \right) \mathrm{d}x\mathrm{d}y = \oint_{\partial\Omega} a(v_x n_x + v_y n_y) \mathrm{d}s - \int_{\Omega} \left(\frac{\partial a}{\partial x} v_x + \frac{\partial a}{\partial y} v_y \right) \mathrm{d}x\mathrm{d}y. \tag{4.12}$$

Now we return to the strong form of the governing partial differential equation (PDE) for planar 2-D, steady state, Fourier heat conduction, (4.5), repeated here with heat source f moved to the left side:

$$\nabla \cdot (\boldsymbol{k} \nabla T) + f = 0, \qquad (x, y) \in \Omega. \tag{4.13}$$

Refer again to Fig. 4.1 for notation. A differentiable but otherwise arbitrary scalar field $w(x, y)$ is then introduced in the domain. Multiplying (4.13) by w and then integrating over area Ω gives

$$\int_{\Omega} [w \nabla \cdot (\boldsymbol{k} \nabla T)] \mathrm{d}x\mathrm{d}y + \int_{\Omega} w f \mathrm{d}x\mathrm{d}y = 0. \tag{4.14}$$

We then apply identity (4.11) to the first integral term on the left side of the equality in (4.14) to produce

$$\int_{\Omega} w [\nabla \cdot (\boldsymbol{k} \nabla T)] \mathrm{d}x\mathrm{d}y = \oint_{\partial\Omega} (w \boldsymbol{k} \nabla T) \cdot \boldsymbol{n} \mathrm{d}s - \int_{\Omega} (\nabla w) \cdot (\boldsymbol{k} \nabla T) \mathrm{d}x\mathrm{d}y. \tag{4.15}$$

Substituting this result back into (4.14), we arrive at the weak form PDE for steady conduction:

$$\int_{\Omega} (\nabla w) \cdot (\boldsymbol{k}\nabla T)\mathrm{d}x\mathrm{d}y = \int_{\Omega} wf\mathrm{d}x\mathrm{d}y + \oint_{\partial\Omega} (w\boldsymbol{k}\nabla T) \cdot \boldsymbol{n}\mathrm{d}s. \tag{4.16}$$

Expressed in terms of scalar components, (4.16) is

$$\begin{aligned} \int_{\Omega} \left[\frac{\partial w}{\partial x} \left(k_{11} \frac{\partial T}{\partial x} + k_{12} \frac{\partial T}{\partial y} \right) + \frac{\partial w}{\partial y} \left(k_{12} \frac{\partial T}{\partial x} + k_{22} \frac{\partial T}{\partial y} \right) \right] \mathrm{d}x\mathrm{d}y \\ = \int_{\Omega} wf\mathrm{d}x\mathrm{d}y - \oint_{\partial\Omega} wq_n\mathrm{d}s, \end{aligned} \tag{4.17}$$

recalling from §4.1 that the heat flux on the boundary (positive for outward energy flow) is, in 2-D Cartesian space,

$$q_n = \boldsymbol{q} \cdot \boldsymbol{n} = q_x n_x + q_y n_y \qquad (s \in \partial\Omega). \tag{4.18}$$

The complete weak form for the problem statement corresponding to Fig. 4.1 seeks the temperature field $T(x,y)$ in domain Ω that satisfies first-order partial differential equation (4.17), given boundary conditions on $\partial\Omega$, imposed heat source f, and conductivity matrix $\boldsymbol{k}$ that may generally depend on position. Notice that the weak form governing equation contains first order partial derivatives of each of w and T, in contrast to the strong form PDE (4.6) that is essentially second order in spatial derivatives of $T(x,y)$. Prior to derivation of the matrix equations for finite elements used in steady heat conduction problems, shape functions for 2-D domains must be introduced.

4.3 Planar 2-D Problems: Interpolation

In 2-D problems, shape functions or interpolation functions are needed in FEA for interpolation of field variables at locations (x,y) differing from discrete nodal coordinates (x_i,y_i). Though the initial context of the current presentation is steady heat conduction problems with interpolation of the sole primary field variable temperature T, we will later use same functions for interpolating scalar components of the displacement field in 2-D elasticity problems. Derivatives of the shape functions will also later be used in computation of the element stiffness matrix for both classes of problems.

Interpolation functions have similar properties of those used for 1-D bars and trusses in §2.3, with the caveat that now both x and y coordinates must be considered in the 2-D situation. By definition, shape functions have the following properties: N_i is nonzero only inside elements sharing/containing node i; N_i has a value of one at node i located at point (x_i,y_i); and at any other node j not located at point (x_i,y_i), N_j has a value of zero. The latter two local properties can be stated succinctly as the identity

$$N_i(x_j, y_j) = \delta_{ij}; \qquad \delta_{ij} = 1 \forall i = j, \quad \delta_{ij} = 0 \forall i \neq j; \tag{4.19}$$

where again δ_{ij} is the Kronecker delta.

In this textbook the focus is on polynomial shape functions, the standard type in FEA. In 2-D, each term in the polynomial should be linearly independent of other terms, and it is desirable, but not essential, that the polynomial be complete. A complete polynomial is a polynomial function containing all terms up to the highest order of completeness. Properties of linear independence and completeness in the context of triangular elements are interpreted visually in Fig. 4.2. Using any term of a given order only once results in linear independence. Including all terms in a full triangle (e.g., within Pascal's triangle) produces a complete polynomial. For example, a satisfactory polynomial for linear elements contains a constant term and one term linear in each of x and y.

Completeness is generally associated with improved convergence properties of a numerical solution, though a detailed discussion of convergence is outside the scope of this textbook. Completeness also enables shape functions to perfectly interpolate solutions corresponding to uniform values of primary field variables (e.g., rigid body modes or constant temperatures) as well as linear variations in primary field variables (e.g., constant strains or uniform temperature gradients).

Some popular element types have shape functions that are incomplete to the highest degree of terms in their corresponding polynomial. However, these elements should still be complete to a lower degree. For example, a 4-node quadrilateral element has highest order two (due to bilinear term involving the product xy) but is complete only to order one because it does not consider terms quadratic in x or y. Visual interpretation using Pascal's triangle for quadrilateral element types is shown in Fig. 4.3.

Now we consider in more detail the shape functions $N_i(x,y)$ for a linear triangular element, with nodal coordinates shown in Fig. 4.4. Since this element contains three nodes, subscript $i = 1,2,3$. For an element with nodes labeled in the counterclockwise sense as shown in Fig. 4.4, the area enclosed by the edges of the triangle is, recalling $\det \boldsymbol{A} = \det(\boldsymbol{A}^{\mathrm{T}})$ for square matrix $\boldsymbol{A}$,

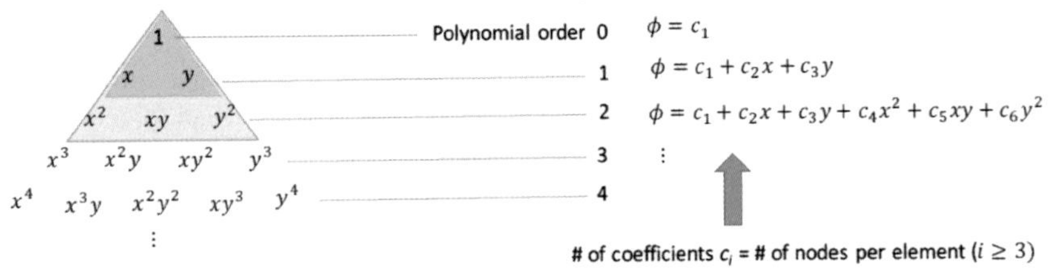

Fig. 4.2 Interpretation of triangular element shape functions via Pascal's triangle

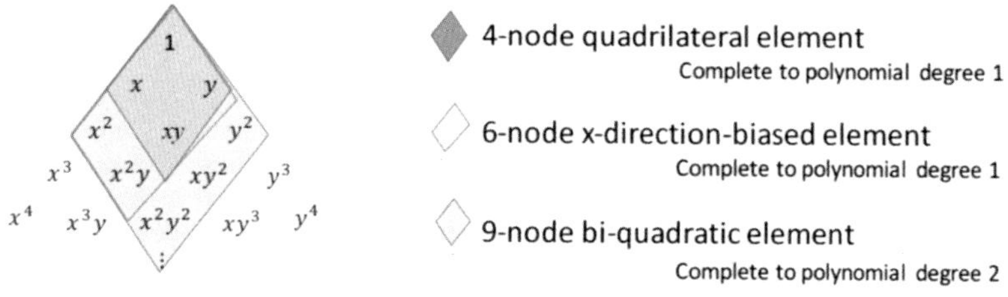

Fig. 4.3 Interpretation of quadrilateral element shape functions via Pascal's triangle

$$
\begin{aligned}
A_e &= \frac{1}{2}(x_1y_2 + x_2y_3 + x_3y_1 - x_1y_3 - x_2y_1 - x_3y_2) \\
&= \frac{1}{2}\det\begin{bmatrix} 1 & x_1 & y_1 \\ 1 & x_2 & y_2 \\ 1 & x_3 & y_3 \end{bmatrix} = \frac{1}{2}\det\begin{bmatrix} 1 & 1 & 1 \\ x_1 & x_2 & x_3 \\ y_1 & y_2 & y_3 \end{bmatrix}.
\end{aligned}
\tag{4.20}
$$

Consulting Fig. 4.2, the polynomial form of the shape function for a linear triangular element is

$$
N_i(x,y) = c_{i1} + c_{i2}x + c_{i3}y, \qquad (i = 1,2,3). \tag{4.21}
$$

The nine coefficients c_{ij} (where subscript $j = 1,2,3$) are determined from the nine constraint equations embedded in (4.19). These conditions are shown graphically in Fig. 4.5 for the three linear shape functions and can be written for this particular element type as

$$
\begin{bmatrix} 1 & x_1 & y_1 \\ 1 & x_2 & y_2 \\ 1 & x_3 & y_3 \end{bmatrix} \begin{Bmatrix} c_{i1} \\ c_{i2} \\ c_{i3} \end{Bmatrix} = \begin{Bmatrix} \delta_{i1} \\ \delta_{i2} \\ \delta_{i3} \end{Bmatrix}. \tag{4.22}
$$

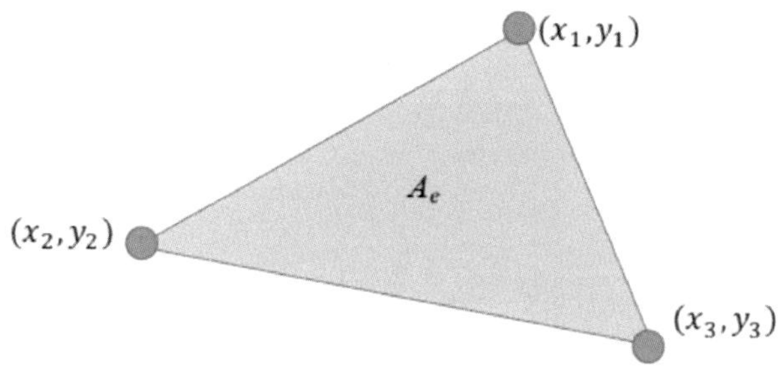

Fig. 4.4 Linear triangular element with nodes at (x_i, y_i) and area A_e

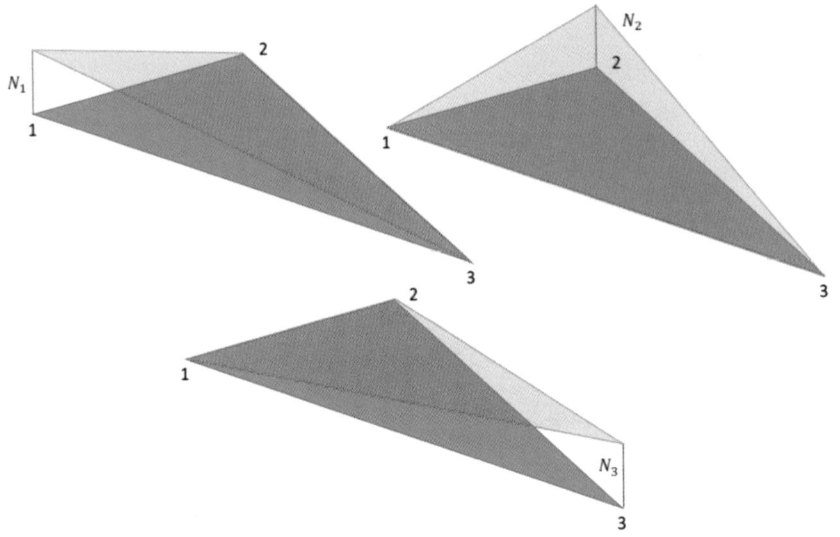

Fig. 4.5 Linear interpolation functions $N_i(x_j, y_j)$ for triangular finite elements

Solution of (4.22) for the c_{ij} coefficients requires use of the inverse of the matrix on its left side:

$$\begin{bmatrix} 1 & x_1 & y_1 \\ 1 & x_2 & y_2 \\ 1 & x_3 & y_3 \end{bmatrix}^{-1} = \frac{1}{2A_e} \begin{bmatrix} \alpha_1 & \alpha_2 & \alpha_3 \\ \beta_1 & \beta_2 & \beta_3 \\ \gamma_1 & \gamma_2 & \gamma_3 \end{bmatrix}, \tag{4.23}$$

where

$$\begin{bmatrix} \alpha_1 & \alpha_2 & \alpha_3 \\ \beta_1 & \beta_2 & \beta_3 \\ \gamma_1 & \gamma_2 & \gamma_3 \end{bmatrix} = \begin{bmatrix} (x_2y_3 - x_3y_2) & (x_3y_1 - x_1y_3) & (x_1y_2 - x_2y_1) \\ y_2 - y_3 & y_3 - y_1 & y_1 - y_2 \\ x_3 - x_2 & x_1 - x_3 & x_2 - x_1 \end{bmatrix}. \tag{4.24}$$

Finally, substituting the determined coefficients c_{ij} back into (4.21) produces the explicit shape functions for the triangular element:

$$N_i(x,y) = \frac{1}{2A_e}(\alpha_i + \beta_i x + \gamma_i y), \qquad (i = 1,2,3), \tag{4.25}$$

where constants α_i, β_i, and γ_i are found from nodal coordinates for the element as shown in (4.24).

Of particular importance is the occurrence of the area of the element A_e in the denominator of each constant entering (4.21) or equivalently in the pre-multiplier in (4.25). Two potential meshing problems emerge from its presence:

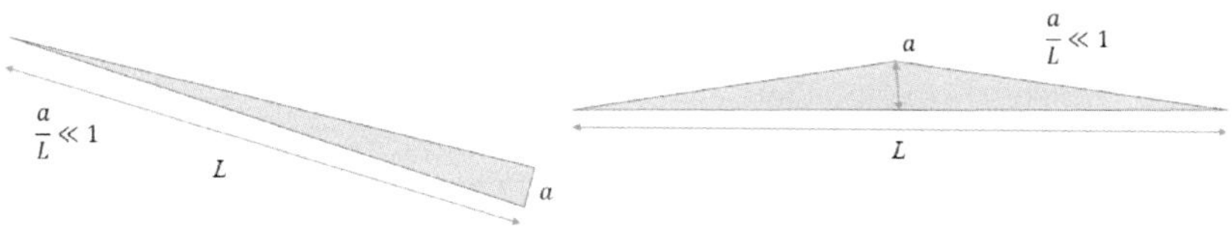

Fig. 4.6 Triangular elements with poor (small) aspect ratios a/L

- **Small aspect ratios**: when any two nodes have spatial coordinates in very close proximity—as occurs when an element is very thin or has very small internal angles as in Fig. 4.6—then $A_e \to 0$, and the shape functions approach an undefined status;
- **Inconsistent node numbering**: when two nodes are switched in terms of local numbering as in Fig. 4.7, then the corresponding two rows are switched in the first matrix in (4.20) and the area changes sign; e.g., A_e becomes negative for clockwise numbering if positive for counter-clockwise numbering, and vice-versa.

The same issues arise when considering quadrilateral elements in 2-D space. Regardless, mesh quality tends to be highest when elements have large aspect ratios, and consistent node numbering is required (counter-clockwise or clockwise, but not both) to avoid instances of elements with negative areas.

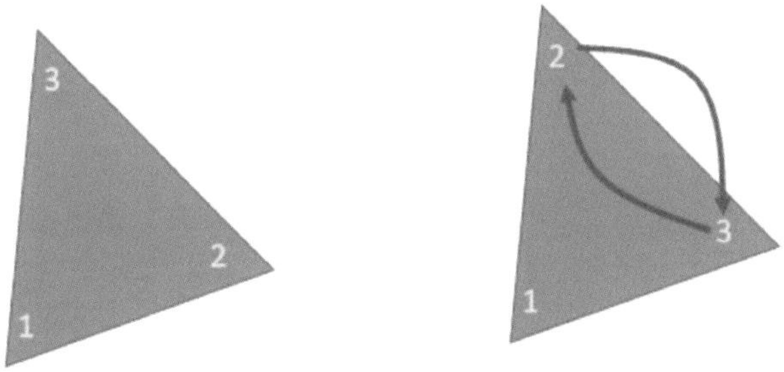

Fig. 4.7 Counter-clockwise node numbering convention (left) and contradictory clockwise numbering scheme (right) for linear triangular elements

4.4 Heat Conduction: Matrix Equations

With the interpolation functions $N_i(x,y)$ now available, the weak form differential equation for steady heat conduction can be used to derive the finite element equilibrium equations for a single element. The weak form PDE is (4.17), which for a domain of integration Ω^e consisting of a single element becomes

$$\begin{aligned}\int_{\Omega^e}\left[\frac{\partial w}{\partial x}\left(k_{11}\frac{\partial T}{\partial x}+k_{12}\frac{\partial T}{\partial y}\right)+\frac{\partial w}{\partial y}\left(k_{12}\frac{\partial T}{\partial x}+k_{22}\frac{\partial T}{\partial y}\right)\right]\mathrm{d}x\mathrm{d}y\\ =\int_{\Omega^e} wf\mathrm{d}x\mathrm{d}y-\oint_{\partial\Omega^e} wq_n\mathrm{d}s.\end{aligned} \tag{4.26}$$

Applying the summation convention for repeated indices, the temperature field $T(x,y)$ and weight function field $w(x,y)$ are discretized as follows:

$$T(x,y)=N_j(x,y)T_j, \qquad w(x,y)=N_i(x,y)w_i, \tag{4.27}$$

where i or j runs from 1 to the total number of nodes or degrees of freedom for element e, and where T_j and w_i are discrete values of temperature and w at node j or i with respective coordinates (x_j,y_j) or (x_i,y_i). Substituting (4.27) into (4.26) and factoring out the constants w_i results in

$$\begin{aligned}w_i\int_{\Omega^e}\left[\frac{\partial N_i}{\partial x}\left(k_{11}\frac{\partial N_j}{\partial x}+k_{12}\frac{\partial N_j}{\partial y}\right)+\frac{\partial N_i}{\partial y}\left(k_{12}\frac{\partial N_j}{\partial x}+k_{22}\frac{\partial N_j}{\partial y}\right)\right]T_j\mathrm{d}x\mathrm{d}y\\ =w_i\int_{\Omega^e} N_i f\mathrm{d}x\mathrm{d}y-w_i\oint_{\partial\Omega^e} N_i q_n\mathrm{d}s.\end{aligned} \tag{4.28}$$

Since the function w and hence its nodal values w_i are arbitrary, it follows that (4.28) represents the following set of scalar equations, one for each value of i:

$$\begin{aligned}\int_{\Omega^e}\left[\frac{\partial N_i}{\partial x}\left(k_{11}\frac{\partial N_j}{\partial x}+k_{12}\frac{\partial N_j}{\partial y}\right)+\frac{\partial N_i}{\partial y}\left(k_{12}\frac{\partial N_j}{\partial x}+k_{22}\frac{\partial N_j}{\partial y}\right)\right]T_j\mathrm{d}x\mathrm{d}y\\ =\int_{\Omega^e} N_i f\mathrm{d}x\mathrm{d}y-\oint_{\partial\Omega^e} N_i q_n\mathrm{d}s.\end{aligned} \tag{4.29}$$

Factoring out the constants T_j from the first integral, (4.29) can be written in vector-matrix form as

$$[\boldsymbol{k}^{(e)}]\{\boldsymbol{T}^{(e)}\}=\{\boldsymbol{f}^{(e)}\}+\{\boldsymbol{Q}^{(e)}\}, \tag{4.30}$$

where $\boldsymbol{T}^{(e)}$ is a column vector of nodal temperatures T_i, and components of the heat source vector $\boldsymbol{f}^{(e)}$ and boundary term $\boldsymbol{Q}^{(e)}$ are defined as

$$f_i^{(e)}=\int_{\Omega^e} N_i(x,y)f(x,y)\mathrm{d}x\mathrm{d}y, \qquad Q_i^{(e)}=-\oint_{\partial\Omega^e} N_i(s)q_n(s)\mathrm{d}s, \tag{4.31}$$

noting that on the element's boundary $\partial\Omega^e$, coordinates can be parametrically described by equations of the form $x = x(s)$ and $y = y(s)$. The boundary contribution in the second equation of (4.31) will be considered in more detail in several examples later in §4.5.

Components of the element stiffness for heat conduction are found via the area integral

$$k_{ij}^e = \int_{\Omega^e} \left[\frac{\partial N_i}{\partial x} \left(k_{11} \frac{\partial N_j}{\partial x} + k_{12} \frac{\partial N_j}{\partial y} \right) + \frac{\partial N_i}{\partial y} \left(k_{12} \frac{\partial N_j}{\partial x} + k_{22} \frac{\partial N_j}{\partial y} \right) \right] \mathrm{d}x\mathrm{d}y. \quad (4.32)$$

Introducing components of the B-matrix for 2-D elements as the following collection of first derivatives, where here j runs from 1 to the total number of nodes n in the element,

$$B_{1j} = \frac{\partial N_j}{\partial x}, \qquad B_{2j} = \frac{\partial N_j}{\partial y}, \quad (4.33)$$

the element stiffness matrix can be expressed succinctly as

$$[\boldsymbol{k}^{(e)}] = \int_{\Omega^e} [\boldsymbol{B}]^{\mathrm{T}} [\boldsymbol{k}] [\boldsymbol{B}] \mathrm{d}x\mathrm{d}y, \quad (4.34)$$

where $\boldsymbol{k}$ is the 2×2 thermal conductivity matrix introduced in (4.3). The size of the element stiffness matrix is $n \times n$, where n is the total number of nodal degrees of freedom for the element (e.g., the number of nodes per element for heat conduction problems).

Derivations to this point in §4.4 apply for any 2-D element, where subscripts $i, j \geq 3$. Now consider linear triangular finite elements with shape functions listed in (4.25). In vector-matrix form,

$$[\boldsymbol{N}(x,y)] = \frac{1}{2A_e} [\alpha_1 + \beta_1 x + \gamma_1 y \quad \alpha_2 + \beta_2 x + \gamma_2 y \quad \alpha_3 + \beta_3 x + \gamma_3 y]. \quad (4.35)$$

The B-matrix for linear triangular elements is the following 2×3 matrix of constants, with transpose:

$$[\boldsymbol{B}] = \frac{1}{2A_e} \begin{bmatrix} \beta_1 & \beta_2 & \beta_3 \\ \gamma_1 & \gamma_2 & \gamma_3 \end{bmatrix}, \qquad [\boldsymbol{B}]^{\mathrm{T}} = \frac{1}{2A_e} \begin{bmatrix} \beta_1 & \gamma_1 \\ \beta_2 & \gamma_2 \\ \beta_3 & \gamma_3 \end{bmatrix}. \quad (4.36)$$

For the particularly simple constitutive model of an isotropic and homogeneous conductor corresponding to the matrix $\boldsymbol{k} = k\mathbf{I}$ in (4.7), where k is the scalar conductivity, the stiffness matrix in (4.34) for a linear triangular element can be obtained analytically by direct integration as

$$k_{ij}^{(e)} = \frac{k}{4A_e} (\beta_i \beta_j + \gamma_i \gamma_j), \qquad (i, j = 1, 2, 3). \quad (4.37)$$

From (4.24), since the terms β_i and γ_i depend only on differences or directional distances between x or y coordinates of nodes in any element and not absolute values of x_i or y_i, entries of (4.37) are independent of rigid body translations of the element.

4.5 Heat Conduction: Boundary Conditions

Natural boundary conditions corresponding to the second term on the right side of the heat conduction equilibrium equation in (4.30), i.e., $\boldsymbol{Q}^{(e)}$, are now revisited in more detail. From (4.18) and (4.31),

$$Q_i^{(e)} = -\oint_{\partial\Omega^e} N_i q_n \mathrm{d}s = -\oint_{\partial\Omega^e} N_i \boldsymbol{q}\cdot\boldsymbol{n}\mathrm{d}s = -\oint_{\partial\Omega^e} N_i(q_x n_x + q_y n_y)\mathrm{d}s, \quad (4.38)$$

where s is a coordinate along the surface $\partial\Omega^e$. Each nodal force component $Q_i^{(e)}$ is thus a weighted integral of the scalar heat flux q_n. In general, $\partial\Omega^e$ may not perfectly match the boundary of a domain $\partial\Omega_q$; for example, a curved boundary is represented by a number of straight line segments (e.g., edges of triangular elements) once discretized into elements. The integral in (4.38) must be evaluated on the union of all element faces $\partial\Omega^e$ that approximate the boundary $\partial\Omega_q$ in Fig. 4.1. Essential boundary conditions require prescription of nodal temperature values T_i at nodes i along $\partial\Omega_T$ and require no further explanation.

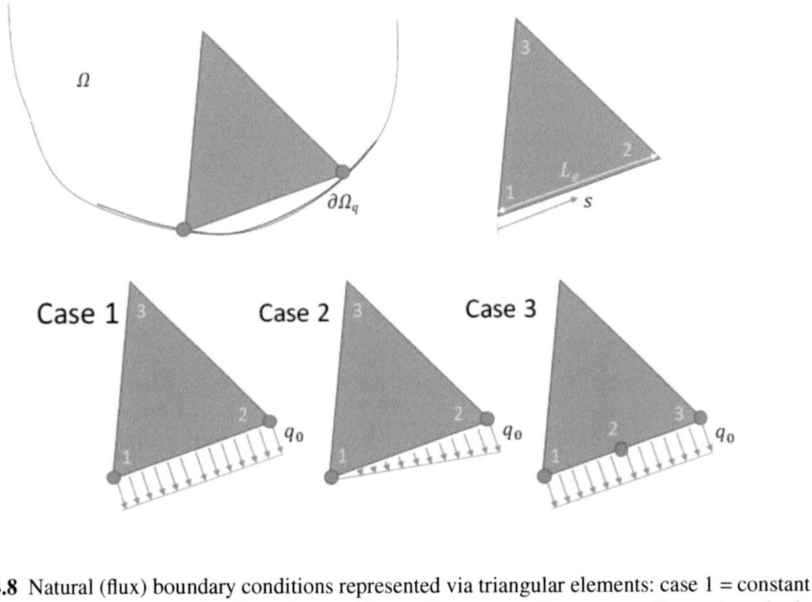

Fig. 4.8 Natural (flux) boundary conditions represented via triangular elements: case 1 = constant flux and linear element, case 2 = linear flux and linear element, case 3 = constant flux and quadratic element

Four examples are considered to demonstrate calculation of $\boldsymbol{Q}^{(e)}$. The first three of these are illustrated in Fig. 4.8, each corresponding to a single triangular finite element. In case 1, flux $q_n = q_0$ is constant over the element face between nodes 1 and 2, where coordinate $s \in [0, L_e]$ parameterizes the edge of length L_e. For a linear triangular element, shape functions N_i are linear along element faces, as is clear from Fig. 4.5:

$$N_1(s) = 1 - s/L_e, \qquad N_2(s) = s/L_e. \tag{4.39}$$

These forms of N_i along an element's edge are identical, with a change in notation, to linear shape functions (2.112) introduced for bar/truss elements in §2.3. For case 1, (4.38) produces

$$Q_1^{(e)} = -\int_0^{L_e} (1 - s/L_e) q_0 \mathrm{d}s = -\frac{1}{2} q_0 L_e, \quad Q_2^{(e)} = -\int_0^{L_e} (s/L_e) q_0 \mathrm{d}s = -\frac{1}{2} q_0 L_e. \tag{4.40}$$

For case 2, a linear triangular element is used again, so (4.39) still applies. However, the prescribed heat flux boundary condition is now the linear function $q_n(s) = q_0 s/L_e$. For case 2, (4.38) therefore results in

$$Q_1^{(e)} = -\int_0^{L_e} \left(1 - \frac{s}{L_e}\right) \frac{q_0 s}{L_e} \mathrm{d}s = -\frac{1}{6} q_0 L_e, \quad Q_2^{(e)} = -\int_0^{L_e} \frac{q_0 s^2}{L_e^2} \mathrm{d}s = -\frac{1}{3} q_0 L_e. \tag{4.41}$$

Now consider case 3, wherein a constant flux $q_n = q_0$ is applied over the face of a quadratic triangular element. Shape functions along the face are quadratic over $s \in [0, L_e]$, with three components since three nodes are involved:

$$N_1(s) = \left(1 - \frac{s}{L_e}\right)\left(1 - \frac{2s}{L_e}\right), \quad N_2(s) = \frac{4s}{L_e}\left(1 - \frac{s}{L_e}\right), \quad N_3(s) = \frac{s}{L_e}\left(\frac{2s}{L_e} - 1\right). \tag{4.42}$$

Then for case 3, thermal forces at the three nodes along $\partial \Omega^e$ are calculated by direct integration as

$$\begin{aligned} Q_1^{(e)} &= -\int_0^{L_e} N_1(s) q_0 \mathrm{d}s = -\frac{1}{6} q_0 L_e, \quad Q_2^{(e)} = -\int_0^{L_e} N_2(s) q_0 \mathrm{d}s = -\frac{2}{3} q_0 L_e, \\ Q_3^{(e)} &= -\int_0^{L_e} N_3(s) q_0 \mathrm{d}s = -\frac{1}{6} q_0 L_e. \end{aligned} \tag{4.43}$$

A fourth and final example considers a domain boundary $\partial \Omega_q$ discretized by edges of two linear triangular elements as shown in Fig. 4.9. Local coordinates $s_{12} \in [0, L_{12}]$ parameterize the edge of the element on the left, which supports a linear flux profile of the form $q_n = q_1 s_{12}/L_{12}$. Local coordinates $s_{23} \in [0, L_{23}]$ parameterize the edge of the element on the right, which supports a constant flux profile of the form $q_n = q_0$. Global node numbers are 1, 2, and 3 as shown. Shape functions for each element are analogous to those in (4.39). At node 1 or 3, only the left or right element contributes to nodal thermal forces via local shape function

$N_1(s_{12})$ or $N_2(s_{23})$, respectively, while node 2 supports the sum of contributions of both elements:

$$
\begin{aligned}
Q_1 &= -\int_0^{L_{12}} \frac{q_1 s_{12}}{L_{12}}\left(1-\frac{s_{12}}{L_{12}}\right)\mathrm{d}s_{12} = -\frac{1}{6}q_1 L_{12},\\
Q_2 &= -\int_0^{L_{12}} \frac{q_1 s_{12}}{L_{12}}\left(\frac{s_{12}}{L_{12}}\right)\mathrm{d}s_{12} - \int_0^{L_{23}} q_0\left(1-\frac{s_{23}}{L_{23}}\right)\mathrm{d}s_{23} = -\frac{1}{3}q_1 L_{12} - \frac{1}{2}q_0 L_{23},\\
Q_3 &= -\int_0^{L_{23}} \frac{q_0 s_{23}}{L_{23}}\mathrm{d}s_{23} = -\frac{1}{2}q_0 L_{23}.
\end{aligned}
\tag{4.44}
$$

Analysis of this two-element case is part of the general assembly process for 2-D steady heat conduction problems as will be described in the next section.

4.6 Heat Conduction: Assembly, Solution, and Post-processing

The assembly process for 2-D heat conduction problems is similar to that for 1-D geometries considered in prior chapters. The global system of equations for steady heat conduction is written as

$$[\boldsymbol{K}]\{\boldsymbol{T}\} = \{\boldsymbol{Q}\} + \{\boldsymbol{f}\} = \{\boldsymbol{F}\}, \tag{4.45}$$

with $\boldsymbol{K}$ the assembled global stiffness matrix, $\boldsymbol{T}$ the column vector of nodal temperatures, $\boldsymbol{Q}$ the column vector of contributions of applied heat fluxes resolved at the nodes, and $\boldsymbol{f}$ the column vector of applied heat sources resolved at the nodes. The total thermal force vector is $\boldsymbol{F}$. If n is the total number of nodes in the domain Ω, then each vector in (4.45) is of length n, and the global stiffness matrix is of dimensions $n \times n$.

At a given node, contributions from all elements sharing that node sum to produce $\boldsymbol{Q}$ and $\boldsymbol{f}$, where element-level contributions are defined by the integrals in

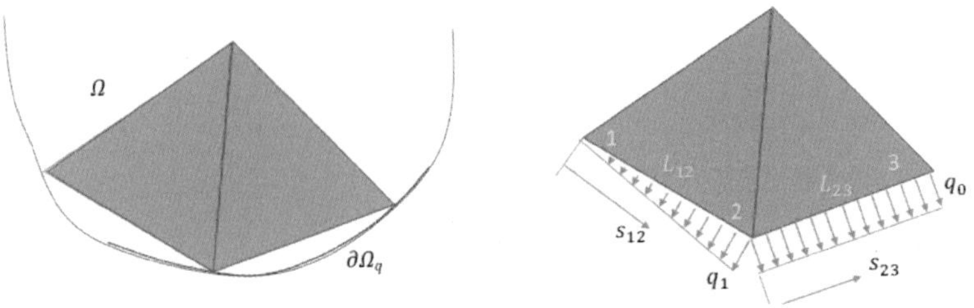

Fig. 4.9 Natural (flux) boundary conditions represented via two linear triangular elements: local coordinates for each element along the piecewise linear discretization of $\partial\Omega_q$ are s_{12} and s_{23}

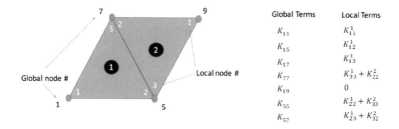

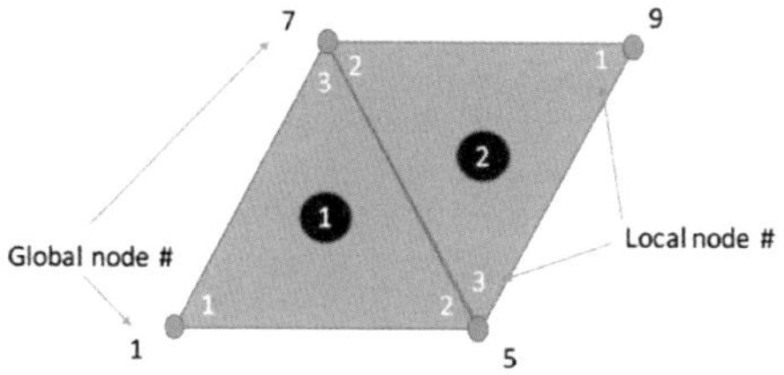

Global Terms	Local Terms
K_{11}	K^1_{11}
K_{15}	K^1_{12}
K_{17}	K^1_{13}
K_{77}	$K^1_{33} + K^2_{22}$
K_{19}	0
K_{55}	$K^1_{22} + K^2_{33}$
K_{57}	$K^1_{23} + K^2_{32}$

Fig. 4.10 Connectivity and representative stiffness matrix contributions for two linear triangular finite elements

(4.31) that involve integration of the shape functions N_i. An example of assembly of the vector $\boldsymbol{Q}$ for two elements was described at the end of §4.5.

The global stiffness matrix $\boldsymbol{K}$ can be assembled, in principle, by consideration of the collected individual element equilibrium equations and mesh connectivity information; i.e., symbolically,

$$[\boldsymbol{K}] = \bigwedge_{e \in \Omega} [\boldsymbol{k}^{(e)}]. \tag{4.46}$$

Such a task, which becomes tedious for a 2-D domain meshed by any significant number of elements, fortunately is performed for the analyst by any modern FEM software. A two-element mesh is considered in Fig. 4.10 to demonstrate a few concepts. First, note that global node numbers are invoked in the global stiffness matrix. Nodes shared by multiple elements produce stiffness contributions from all shared elements, but the stiffness matrix entry corresponding to a DOF not supported by an element contains no contribution from that element. The latter point is demonstrated by entry $K_{19} = 0$ in Fig. 4.10.

Internal and external forces/fluxes are considered in more detail for a domain discretized by many elements in Fig. 4.11. Internal forces balance along inter-element boundaries; for equilibrium,

$$q_n^e + q_n^{e+1} = 0, \tag{4.47}$$

where each term on the left is the scalar heat flux internally supported along an element's edge as shown in the figure. Elements with faces resolving $\partial\Omega_q$ may have nonzero imposed fluxes ($q_n \neq 0$) on such faces. All other element faces along $\partial\Omega \setminus \partial\Omega_q \cup \partial\Omega_T$, i.e., faces where no flux or temperature is prescribed explicitly, support by default a null flux boundary condition $q_n = 0$.

The remainder of the analysis of a 2-D steady heat conduction problem using FEA follows the same general procedure as those for truss elements (Chapter 2) and beam elements (Chapter 3). Specifically, the remaining steps in a static 2-D heat conduction analysis are

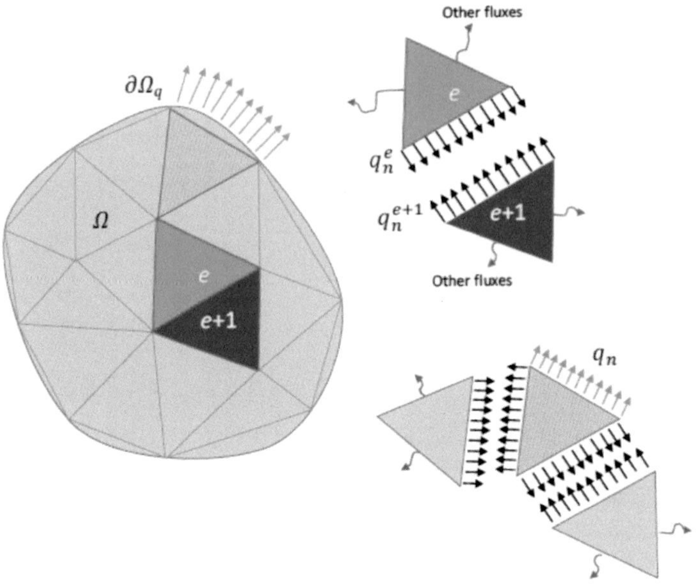

Fig. 4.11 Internal versus external forces (fluxes) for a domain Ω discretized by many triangular elements; here, internal fluxes obey $q_n^e + q_n^{e+1} = 0$

- **Boundary Conditions**: manipulate the global equations to account for essential boundary conditions on primary solution variable T, moving contributions of nonzero imposed generalized displacements T_i on $\partial\Omega_T$ to the right side of the global system of equations in (4.45);
- **Solve**: solve the condensed global system $[\boldsymbol{K}]\{\boldsymbol{T}\} = \{\boldsymbol{F}\}$ where boundary conditions have been imposed, such that the condensed global stiffness matrix $[\boldsymbol{K}]$ is non-singular and thus invertible;
- **Post-process**: compute secondary nodal and elemental quantities by substituting primary solution variables ($\boldsymbol{T}$) into element equations.

A few remarks on post-processing are in order for 2-D heat conduction problems. Consider determination of the heat flux vector $\boldsymbol{q}^{(e)}$ for an element. Noting Fourier's law $\boldsymbol{q} = -\boldsymbol{k}\nabla T$ of (4.4) holds within any element, the heat flux vector for that element can be computed from the primary solution field $\boldsymbol{T}^{(e)}$ referred to local nodes of that element as

$$\{\boldsymbol{q}^{(e)}\} = -[\boldsymbol{k}]\{\nabla T\}^{(e)} = -[\boldsymbol{k}]\nabla([\boldsymbol{N}]\{\boldsymbol{T}^{(e)}\}) = -[\boldsymbol{k}][\boldsymbol{B}]\{\boldsymbol{T}^{(e)}\}, \tag{4.48}$$

where conductivity matrix $\boldsymbol{k}$ and shape function derivative matrix $\boldsymbol{B}$ also depend on the element number. For linear triangular elements, since shape functions are linear, temperature gradients are constant within each element. Thus, for heterogeneous temperature profiles, heat fluxes determined in this way will usually be discontinuous along edges between shared elements. For this reason, nodal values of the heat flux components are often averaged from neighboring element values as a post-processing option in FE software packages.

4.7 Planar Elasticity: Governing Equations

Planar elasticity problems are analyzed via finite elements in a similar manner as planar heat conduction problems, since in both instances solution variables generally depend on Cartesian x and y coordinates. The main source of additional overhead in the former is that the primary solution variable is a vector field (the displacement) rather than a scalar field (the temperature) as in the latter.

Governing equations from continuum mechanics and elasticity theory are presented next. Only essential aspects are discussed; for more in-depth treatments, the reader is referred to books on continuum mechanics and elasticity, e.g., [1, 3, 4]. The content of this textbook is restricted to classical linear elasticity, wherein displacements and displacement gradients are presumed small in magnitude and wherein stress is a linear function of strain.

In planar 2-D solid mechanics problems, the displacement vector $\boldsymbol{u}$ consists of two in-plane components:

$$\{\boldsymbol{u}(x,y)\} = \begin{Bmatrix} u_1(x_1,x_2) \\ u_2(x_1,x_2) \end{Bmatrix} = \begin{Bmatrix} u_x(x,y) \\ u_y(x,y) \end{Bmatrix} = \begin{Bmatrix} u(x,y) \\ \upsilon(x,y) \end{Bmatrix}, \tag{4.49}$$

where the (u, υ) notation will be used most frequently herein. The symmetric strain tensor is defined as

$$\boldsymbol{\varepsilon} = \frac{1}{2}[\nabla\boldsymbol{u} + (\nabla\boldsymbol{u})^{\mathrm{T}}], \tag{4.50}$$

or in index notation as

$$\varepsilon_{ij} = \frac{1}{2}\left(\frac{\partial u_i}{\partial x_j} + \frac{\partial u_j}{\partial x_i}\right). \tag{4.51}$$

Compatibility conditions that ensure that a displacement vector exists producing a given strain field impose restrictions on derivatives of strain. In planar 2-D problems, in-plane strain components are often written in vector form as

$$\begin{Bmatrix} \varepsilon_{xx} \\ \varepsilon_{yy} \\ \gamma_{xy} \end{Bmatrix} = \begin{Bmatrix} \varepsilon_{xx} \\ \varepsilon_{yy} \\ 2\varepsilon_{xy} \end{Bmatrix} = \begin{Bmatrix} \varepsilon_{11} \\ \varepsilon_{22} \\ 2\varepsilon_{12} \end{Bmatrix} = \begin{Bmatrix} \partial u/\partial x \\ \partial \upsilon/\partial y \\ \partial u/\partial y + \partial \upsilon/\partial x \end{Bmatrix}. \tag{4.52}$$

For planar 2-D problems, compatibility of a given strain field demands that the following identity hold for second derivatives of strain components [1, 3]:

$$\frac{\partial^2 \varepsilon_{xx}}{\partial y^2} + \frac{\partial^2 \varepsilon_{yy}}{\partial x^2} = \frac{\partial^2 \gamma_{xy}}{\partial x \partial y}. \tag{4.53}$$

The conservation law to be solved in a static continuum mechanics problem is the local balance of linear momentum:

$$\nabla \cdot \boldsymbol{\sigma} + \boldsymbol{f} = 0. \tag{4.54}$$

The stress tensor is $\boldsymbol{\sigma}$, and the body force vector (force per unit volume) is $\boldsymbol{f}$. A common example of a body force is that due to gravity. Note that for (quasi-)static problems, no inertial terms involving the product of mass density and local acceleration are included. The local balance of angular momentum requires that the stress tensor be symmetric, i.e., $\sigma_{ij} = \sigma_{ji}$, and is not solved explicitly.

In index notation in a general 3-D problem, (4.54) is the following three independent equations ($i = 1,2,3$):

$$\frac{\partial \sigma_{ij}}{\partial x_j} + f_i = 0. \tag{4.55}$$

For 2-D planar problems, (4.54) consists of two independent equations:

$$\frac{\partial \sigma_{xx}(x,y)}{\partial x} + \frac{\partial \sigma_{xy}(x,y)}{\partial y} + f_x(x,y) = 0, \tag{4.56}$$

$$\frac{\partial \sigma_{xy}(x,y)}{\partial x} + \frac{\partial \sigma_{yy}(x,y)}{\partial y} + f_y(x,y) = 0. \tag{4.57}$$

The constitutive equations of linear elasticity, which may be isotropic or more generally anisotropic, are invoked. Stress is linearly related to strain via the matrix/tensor of elastic constants $\boldsymbol{C}$. In full index notation for a general 3-D problem, the linear stress-strain law is

$$\sigma_{ij} = C_{ijkl}\varepsilon_{kl}, \qquad (i,j,k,l = 1,2,3). \tag{4.58}$$

An equivalent constitutive equation is often written using Voigt notation as follows: [3, 4]:

$$\{\boldsymbol{\sigma}\} = [\boldsymbol{C}]\{\boldsymbol{\varepsilon}\} \leftrightarrow \sigma_I = C_{IJ}\varepsilon_J, \qquad (I,J = 1,2,\dots 6). \tag{4.59}$$

In (4.59), $\boldsymbol{C}$ is a symmetric 6×6 matrix relating column vectors of stress and strain. For the most general class of anisotropic material, $\boldsymbol{C}$ consists of 21 independent constants; material symmetries reduce this number to a minimum of 2 elastic constants for isotropy. The strain energy density W, in dimensions of energy per unit volume, is a quadratic form with first and second derivatives with respect to strain given as

$$W = \frac{1}{2}C_{ijkl}\varepsilon_{ij}\varepsilon_{kl} \Rightarrow \sigma_{ij} = \frac{\partial W}{\partial \varepsilon_{ij}}, \qquad C_{ijkl} = \frac{\partial \sigma_{ij}}{\partial \varepsilon_{kl}} = \frac{\partial^2 W}{\partial \varepsilon_{ij}\partial \varepsilon_{kl}}. \tag{4.60}$$

It follows that the elasticity tensor has the symmetries $C_{ijkl} = C_{jikl}$, $C_{ijkl} = C_{ijlk}$, and $C_{klij} = C_{ijkl}$ that enable a reduction from $3^4 = 81$ to the 21 independent components mentioned above. Applying such symmetries, stress and strain energy density can be expressed in terms of displacement gradients as

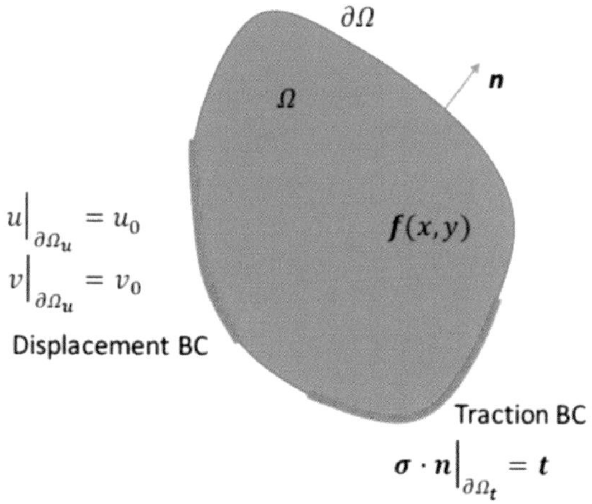

Fig. 4.12 Domain Ω for solid mechanics (stress) analysis in two dimensions with boundary conditions

$$\sigma_{ij} = C_{ijkl}\frac{\partial u_k}{\partial x_l} = C_{ijkl}\frac{\partial u_l}{\partial x_k}, \qquad W = \frac{1}{2}C_{ijkl}\frac{\partial u_i}{\partial x_j}\frac{\partial u_k}{\partial x_l}. \tag{4.61}$$

As shown in Fig. 4.12, we consider a 2-D domain Ω with boundary $\partial\Omega$, where Cartesian coordinates x and y cover the domain. The unit outward normal vector to Ω is $\boldsymbol{n}(x,y)$. Boundary conditions on $\partial\Omega$ are of two possible types:

- **Essential**: displacements (u, v) prescribed on $\partial\Omega_u$;
- **Natural**: traction vector components (t_x, t_y) prescribed on $\partial\Omega_t$.

The mechanical traction vector on the boundary is defined as

$$\boldsymbol{t} = \boldsymbol{\sigma}\cdot\boldsymbol{n} = \boldsymbol{n}\cdot\boldsymbol{\sigma} \leftrightarrow t_i = \sigma_{ij}n_j = n_j\sigma_{ji}. \tag{4.62}$$

It is also possible to invoke mixed boundary conditions, e.g., imposed (u, t_y) or (t_x, v), but imposition of both entries of a work conjugate pair (u, t_x) or (v, t_y) at the same location on $\partial\Omega$ is physically forbidden.

More specific constitutive laws apply for plane stress or plane strain boundary conditions as discussed next.

4.7.1 Plane Stress

Plane stress conditions physically correspond to thin, planar-type bodies. All out-of-plane stress components are defined as zero, such that for the present coordinate system,

$$\sigma_{zz} = \sigma_{xz} = \sigma_{yz} = 0, \qquad (\text{generally, } \varepsilon_{zz} \neq 0). \tag{4.63}$$

Strain $\varepsilon_{zz} = \partial u_z / \partial z$ is generally nonzero since the thin body is free to expand or contract out-of-plane. For a linear elastic material with orthotropic or higher symmetry, planar stress and strain components are related by

$$\begin{Bmatrix} \sigma_{xx} \\ \sigma_{yy} \\ \sigma_{xy} \end{Bmatrix} = \begin{bmatrix} C_{11} & C_{12} & 0 \\ C_{12} & C_{22} & 0 \\ 0 & 0 & C_{66} \end{bmatrix} \begin{Bmatrix} \varepsilon_{xx} \\ \varepsilon_{yy} \\ \gamma_{xy} \end{Bmatrix}. \tag{4.64}$$

For a linear elastic material with isotropic symmetry, Young's modulus E, and Poisson's ratio ν, (4.64) reduces to

$$\begin{Bmatrix} \sigma_{xx} \\ \sigma_{yy} \\ \sigma_{xy} \end{Bmatrix} = \frac{E}{1-\nu^2} \begin{bmatrix} 1 & \nu & 0 \\ \nu & 1 & 0 \\ 0 & 0 & \frac{1}{2}(1-\nu) \end{bmatrix} \begin{Bmatrix} \varepsilon_{xx} \\ \varepsilon_{yy} \\ \gamma_{xy} \end{Bmatrix}. \tag{4.65}$$

4.7.2 Plane Strain

Plane strain conditions physically correspond to long geometries (in the z-direction) with uniform cross-sections. All out-of-plane strains are defined to vanish, such that for the present coordinate system,

$$\varepsilon_{zz} = \varepsilon_{xz} = \varepsilon_{yz} = 0, \qquad (\text{generally, } \sigma_{zz} \neq 0). \tag{4.66}$$

Stress σ_{zz} is generally nonzero since forces may be needed to prevent the body from expanding or contracting out-of-plane. For a linear elastic material with orthotropic or higher symmetry, planar stress and strain components are related by an equation identical in appearance to (4.64), but components C_{ij} (where $i,j = 1,2,6$) differ for plane stress and plane strain loading. For a linear elastic material with isotropic symmetry, Young's modulus E, and Poisson's ratio ν, (4.64) reduces for plane strain conditions to

$$\begin{Bmatrix} \sigma_{xx} \\ \sigma_{yy} \\ \sigma_{xy} \end{Bmatrix} = \frac{E}{(1+\nu)(1-2\nu)} \begin{bmatrix} 1-\nu & \nu & 0 \\ \nu & 1-\nu & 0 \\ 0 & 0 & \frac{1}{2}(1-2\nu) \end{bmatrix} \begin{Bmatrix} \varepsilon_{xx} \\ \varepsilon_{yy} \\ \gamma_{xy} \end{Bmatrix}. \tag{4.67}$$

4.7.3 Strong Form

Returning now to general planar 2-D elasticity problems, i.e., either plane stress or plane strain problems, the strong form PDEs are derived as follows. The constitutive equations in (4.64)—noting $C_{ij}(x,y)$ depend on the choice of plane stress or plane strain and generally also are functions of position—are substituted into equilibrium equations (4.56) and (4.57) to produce, respectively,

$$\frac{\partial}{\partial x}(C_{11}\varepsilon_{xx}+C_{12}\varepsilon_{yy})+\frac{\partial}{\partial y}(C_{66}\gamma_{xy})+f_x=0, \tag{4.68}$$

$$\frac{\partial}{\partial y}(C_{22}\varepsilon_{yy}+C_{12}\varepsilon_{xx})+\frac{\partial}{\partial x}(C_{66}\gamma_{xy})+f_y=0. \tag{4.69}$$

Now substituting the strain definitions in (4.52), we arrive at the following system of two coupled PDEs, each generally of second order, in displacement components $u(x,y)$ and $\upsilon(x,y)$:

$$\frac{\partial}{\partial x}\left(C_{11}\frac{\partial u}{\partial x}+C_{12}\frac{\partial \upsilon}{\partial y}\right)+\frac{\partial}{\partial y}\left[C_{66}\left(\frac{\partial u}{\partial y}+\frac{\partial \upsilon}{\partial x}\right)\right]+f_x=0, \tag{4.70}$$

$$\frac{\partial}{\partial y}\left(C_{22}\frac{\partial \upsilon}{\partial y}+C_{12}\frac{\partial u}{\partial x}\right)+\frac{\partial}{\partial x}\left[C_{66}\left(\frac{\partial u}{\partial y}+\frac{\partial \upsilon}{\partial x}\right)\right]+f_y=0. \tag{4.71}$$

These equations can be simplified, of course, when C_{ij} does not depend on x or y (i.e., a homogeneous material) and/or when isotropic material symmetry relation (4.65) or (4.67) holds. The complete strong form for the problem statement corresponding to Fig. 4.12 seeks the displacement field $[u(x,y),\upsilon(x,y)]$ in domain Ω that satisfies second-order partial differential equations (4.70) and (4.71), given boundary conditions on $\partial\Omega$, imposed body force field $\boldsymbol{f}$, and elasticity matrix $\boldsymbol{C}$ that may generally depend on position.

4.8 Planar Elasticity: Weak Form

The weak form of the problem statement for planar 2-D elasticity problems is derived from the strong form following a procedure analogous to that for heat conduction problems considered already in §4.2. Recall that the present derivation is restricted to a quasi-static and linear elastic response, where the entire problem domain Ω is shown in Fig. 4.12. As noted in §4.7.1 and §4.7.2 for plane stress and plane strain, respectively, symmetry of the elasticity tensor used in subsequent derivations also requires that the solid be of at least orthotropic symmetry, with up to three independent in-plane elastic constants.

The derivation of the weak form PDEs begins with introduction of a differentiable but otherwise arbitrary vector-valued weight function $\boldsymbol{w}(x,y)$, with two components for 2-D analyses:

$$\{\boldsymbol{w}(\boldsymbol{x})\}=\begin{Bmatrix} w_1(x_1,x_2) \\ w_2(x_1,x_2) \end{Bmatrix}=\begin{Bmatrix} w_x(x,y) \\ w_y(x,y) \end{Bmatrix}. \tag{4.72}$$

The derivation proceeds most efficiently when coordinate-free tensor notation is invoked. Taking the scalar product of $\boldsymbol{w}$ with linear momentum balance (4.54) and integrating over the problem domain produces

$$\int_{\Omega} \boldsymbol{w} \cdot (\nabla \cdot \boldsymbol{\sigma}) \mathrm{d}\Omega + \int_{\Omega} \boldsymbol{w} \cdot \boldsymbol{f} \mathrm{d}\Omega = 0. \tag{4.73}$$

Integrating the first term on the left by parts with application of the divergence theorem, i.e., the integral theorem of Gauss discussed in §4.2, gives

$$\begin{aligned} \int_{\Omega} \boldsymbol{w} \cdot (\nabla \cdot \boldsymbol{\sigma}) \mathrm{d}\Omega &= \int_{\Omega} \nabla \cdot (\boldsymbol{w} \cdot \boldsymbol{\sigma}) \mathrm{d}\Omega - \int_{\Omega} (\nabla \boldsymbol{w}) : \boldsymbol{\sigma} \mathrm{d}\Omega \\ &= \oint_{\partial\Omega} \boldsymbol{w} \cdot \boldsymbol{\sigma} \cdot \boldsymbol{n} \, \mathrm{d}\partial\Omega - \int_{\Omega} (\nabla \boldsymbol{w}) : \boldsymbol{\sigma} \mathrm{d}\Omega. \end{aligned} \tag{4.74}$$

The scalar product of two second-order tensors is, in index notation,

$$(\nabla \boldsymbol{w}) : \boldsymbol{\sigma} = \frac{\partial w_i}{\partial x_j} \sigma_{ij}. \tag{4.75}$$

Substituting (4.74) into (4.73) and using the definition of the traction vector from (4.62), we arrive at the following integral PDE:

$$\int_{\Omega} (\nabla \boldsymbol{w}) : \boldsymbol{\sigma} \mathrm{d}\Omega - \oint_{\partial\Omega} \boldsymbol{w} \cdot \boldsymbol{t} \mathrm{d}\partial\Omega - \int_{\Omega} \boldsymbol{w} \cdot \boldsymbol{f} \mathrm{d}\Omega = 0. \tag{4.76}$$

Now invoking linear elasticity constitutive equation $\boldsymbol{\sigma} = \boldsymbol{C} : \boldsymbol{\varepsilon} = \boldsymbol{C} : (\nabla \boldsymbol{u})$ from (4.61), tensor relation (4.76) can be expressed as

$$\int_{\Omega} (\nabla \boldsymbol{w}) : \boldsymbol{C} : (\nabla \boldsymbol{u}) \mathrm{d}\Omega - \oint_{\partial\Omega} \boldsymbol{w} \cdot \boldsymbol{t} \mathrm{d}\partial\Omega - \int_{\Omega} \boldsymbol{w} \cdot \boldsymbol{f} \mathrm{d}\Omega = 0. \tag{4.77}$$

In indicial notation, which also applies for the general 3-D case when $(i, j, k, l = 1, 2, 3)$,

$$\int_{\Omega} \frac{\partial w_i}{\partial x_j} C_{ijkl} \frac{\partial u_k}{\partial x_l} \mathrm{d}\Omega - \oint_{\partial\Omega} w_i t_i \mathrm{d}\partial\Omega - \int_{\Omega} w_i f_i \mathrm{d}\Omega = 0. \tag{4.78}$$

Equations (4.77) and (4.78) are scalar representations of the weak form PDE for quasi-static linear elasticity. For 2-D problems, a weak form problem statement corresponding to Fig. 4.12 seeks the displacement field $\boldsymbol{u}(\boldsymbol{x}) = [u(x,y), \upsilon(x,y)]$ in domain Ω that satisfies these partial differential equations (4.77) or (4.78), given boundary conditions on $\partial\Omega$, imposed body force field $\boldsymbol{f}$, and elasticity matrix $\boldsymbol{C}$ that may generally depend on position. Notice that the governing PDE contains only first derivatives of $\boldsymbol{u}$ and $\boldsymbol{w}$ with respect to x and y, in contrast to the strong form PDEs in (4.70) and (4.71) that contain second derivatives of $\boldsymbol{u}$. The single scalar equation for the weak form derived above will later be parsed into multiple or a vector-valued set of equilibrium equations upon introduction of shape functions and subsequent factorization.

4.9 Planar Elasticity: Matrix Equations

Derivation of the finite element equilibrium equations for planar elasticity is very similar to derivation of the analogous equilibrium equations for planar heat conduction. Therefore, the steps described next parallel those of §4.4. The same general classes of 2-D interpolation functions $N_i(x,y)$ introduced in §4.3 are used here, as are particular shape functions for linear triangular elements in some examples that follow.

The weak form differential equations for quasi-static linear elasticity are used to derive the finite element equilibrium equation for a single element. The main source of increased complexity over the heat conduction derivation is that the primary solution variable is a vector field [displacement $\boldsymbol{u}(x,y)$] rather than a scalar field [temperature $T(x,y)$]. In the present case of elasticity, the number of degrees of freedom per node is thus two $(u_x,u_y) = (u,\upsilon)$ rather than one (T) as in heat conduction.

A weak form PDE for linear elasticity is (4.78), which for a domain of integration Ω^e consisting of a single element becomes

$$\int_{\Omega^e} \frac{\partial w_i}{\partial x_j} C_{ijkl} \frac{\partial u_k}{\partial x_l} \mathrm{d}\Omega = \int_{\Omega^e} w_i f_i \mathrm{d}\Omega + \oint_{\partial\Omega^e} w_i t_i \mathrm{d}\partial\Omega. \tag{4.79}$$

Applying the summation convention over repeated indices, displacement fields $u(x,y)$ and $\upsilon(x,y)$ and weight function field $w_x(x,y)$ and $w_y(x,y)$ are discretized as follows:

$$\begin{Bmatrix} u(x,y) \\ \upsilon(x,y) \end{Bmatrix} = \begin{Bmatrix} N_j(x,y)u_j \\ N_j(x,y)\upsilon_j \end{Bmatrix}, \qquad \begin{Bmatrix} w_x(x,y) \\ w_y(x,y) \end{Bmatrix} = \begin{Bmatrix} N_i(x,y)w_{xi} \\ N_i(x,y)w_{yi} \end{Bmatrix}. \tag{4.80}$$

Subscripts i or j span values of 1 to the total number of nodes n in element e, and u_j and υ_j are nodal values of displacements at node j with Cartesian coordinates (x_j,y_j). Also, (w_{xi},w_{yi}) are discrete values of (w_x,w_y) at node i with coordinates (x_i,y_i). These discrete values will later be alternatively written as w_{ki}, where k corresponds to x or y for planar 2-D problems. The discretized displacement functions can be written more concisely as

$$\{\boldsymbol{u}\} = \begin{Bmatrix} u \\ \upsilon \end{Bmatrix} = [\boldsymbol{N}]\{\boldsymbol{u}^{(e)}\} = \begin{bmatrix} N_1 & 0 & N_2 & 0 & N_3 & 0 & \dots \\ 0 & N_1 & 0 & N_2 & 0 & N_3 & \dots \end{bmatrix} \begin{Bmatrix} u_1 \\ \upsilon_1 \\ u_2 \\ \upsilon_2 \\ u_3 \\ \upsilon_3 \\ \vdots \end{Bmatrix}, \tag{4.81}$$

with $\boldsymbol{N}$ a $2 \times 2n$ matrix of shape functions padded with zeros as indicated. The B-matrix for 2-D planar elasticity is defined as the following $3 \times 2n$ matrix of shape function derivatives:

$$[\boldsymbol{B}] = \begin{bmatrix} \frac{\partial N_1}{\partial x} & 0 & \frac{\partial N_2}{\partial x} & 0 & \frac{\partial N_3}{\partial x} & 0 & \cdots \\ 0 & \frac{\partial N_1}{\partial y} & 0 & \frac{\partial N_2}{\partial y} & 0 & \frac{\partial N_3}{\partial y} & \cdots \\ \frac{\partial N_1}{\partial y} & \frac{\partial N_1}{\partial x} & \frac{\partial N_2}{\partial y} & \frac{\partial N_2}{\partial x} & \frac{\partial N_3}{\partial y} & \frac{\partial N_3}{\partial x} & \cdots \end{bmatrix}. \tag{4.82}$$

This definition enables computation of the strain vector in (4.52) at any location within the element e from its vector of nodal displacements $\boldsymbol{u}^{(e)}$ as

$$\{\boldsymbol{\varepsilon}^{(e)}(x,y)\} = \begin{Bmatrix} \varepsilon_{xx}(x,y) \\ \varepsilon_{yy}(x,y) \\ \gamma_{xy}(x,y) \end{Bmatrix} = [\boldsymbol{B}(x,y)]\{\boldsymbol{u}^{(e)}\}. \tag{4.83}$$

We now return to the weak form equation. Substituting (4.80) into (4.79) and factoring out the constants w_{im} [recalling $w_i(x,y) = N_m(x,y)w_{im}$ with i referring to x or y and m referring to a local node number] results in

$$w_{im}\int_{\Omega^e}\left[\frac{\partial N_m}{\partial x_j}C_{ijkl}\frac{\partial u_k}{\partial x_l}\right]\mathrm{d}x\mathrm{d}y = w_{im}\int_{\Omega^e} N_m f_i \mathrm{d}x\mathrm{d}y + w_{im}\oint_{\partial\Omega^e} N_m t_i \mathrm{d}s. \tag{4.84}$$

Since $[w_x(x,y), w_y(x,y)]$ and hence w_{im} are arbitrary, it follows that (4.84) leads to the following set of $m \times i = 2n$ scalar equations, two for each value of m or two for each node:

$$\int_{\Omega^e}\left[\frac{\partial N_m}{\partial x_j}C_{ijkl}\frac{\partial u_k}{\partial x_l}\right]\mathrm{d}x\mathrm{d}y = \int_{\Omega^e} N_m f_i \mathrm{d}x\mathrm{d}y + \oint_{\partial\Omega^e} N_m t_i \mathrm{d}s. \tag{4.85}$$

By taking advantage of the symmetry of elastic moduli C_{ijkl} and using (4.83), for 2-D planar elasticity, this equation can be written more concisely in vector-matrix form as

$$\left[\int_{\Omega^e}[\boldsymbol{B}]^{\mathrm{T}}[\boldsymbol{C}][\boldsymbol{B}]\mathrm{d}x\mathrm{d}y\right]\{\boldsymbol{u}^{(e)}\} = \int_{\Omega^e}[\boldsymbol{N}]^{\mathrm{T}}\{\boldsymbol{f}\}\mathrm{d}x\mathrm{d}y + \oint_{\partial\Omega^e}[\boldsymbol{N}]^{\mathrm{T}}\{\boldsymbol{t}\}\mathrm{d}s. \tag{4.86}$$

Nodal displacements $\boldsymbol{u}^{(e)}$ have been factored from the first term on left side since they are constant over the element domain Ω_e. In symbolic form, (4.86) is expressed as

$$[\boldsymbol{k}^{(e)}]\{\boldsymbol{u}^{(e)}\} = \{\boldsymbol{f}^{(e)}\} + \{\boldsymbol{Q}^{(e)}\} = \{\boldsymbol{F}^{(e)}\}, \tag{4.87}$$

where $\boldsymbol{k}^{(e)}$ is the $2n \times 2n$ element stiffness matrix, $\boldsymbol{f}^{(e)}$ is the body force vector resolved at the nodes, and $\boldsymbol{Q}^{(e)}$ is the contribution of natural boundary conditions (i.e., applied tractions) resolved at the nodes. The latter two column vectors, and the their sum, the total load vector $\boldsymbol{F}^{(e)}$, are each of length $2n$. Describing quantities entering (4.86) for planar 2-D elasticity, $\boldsymbol{C}$ is a symmetric 3×3 matrix of the form in (4.64), $\boldsymbol{B}$ is a $3 \times 2n$ matrix of the form in (4.82), $\boldsymbol{N}$ is a $2 \times 2n$ matrix of the form in (4.81), $\boldsymbol{f}$ is simply the body force vector with components (f_x, f_y), and $\boldsymbol{t}$ is simply the traction vector of (4.62) with components (t_x, t_y).

Derivations for finite element equilibrium equations of linear elasticity up to this point in §4.9 apply for any planar 2-D element referred to Cartesian coordinates. Now consider linear triangular finite elements with shape functions listed in (4.25). In vector-matrix form, interpolation functions of (4.81) and their derivatives in (4.82) are

$$[\boldsymbol{N}(x,y)]^{\mathrm{T}} = \frac{1}{2A_e}\begin{bmatrix} \alpha_1+\beta_1 x+\gamma_1 y & 0 \\ 0 & \alpha_1+\beta_1 x+\gamma_1 y \\ \alpha_2+\beta_2 x+\gamma_2 y & 0 \\ 0 & \alpha_2+\beta_2 x+\gamma_2 y \\ \alpha_3+\beta_3 x+\gamma_3 y & 0 \\ 0 & \alpha_3+\beta_3 x+\gamma_3 y \end{bmatrix}; \tag{4.88}$$

$$[\boldsymbol{B}] = \frac{1}{2A_e}\begin{bmatrix} \beta_1 & 0 & \beta_2 & 0 & \beta_3 & 0 \\ 0 & \gamma_1 & 0 & \gamma_2 & 0 & \gamma_3 \\ \gamma_1 & \beta_1 & \gamma_2 & \beta_2 & \gamma_3 & \beta_3 \end{bmatrix}, \qquad [\boldsymbol{B}]^{\mathrm{T}} = \frac{1}{2A_e}\begin{bmatrix} \beta_1 & 0 & \gamma_1 \\ 0 & \gamma_1 & \beta_1 \\ \beta_2 & 0 & \gamma_2 \\ 0 & \gamma_2 & \beta_2 \\ \beta_3 & 0 & \gamma_3 \\ 0 & \gamma_3 & \beta_3 \end{bmatrix}. \tag{4.89}$$

Notice that $\boldsymbol{B}$ depends only on the area of the element A_e and the scalar coefficients γ_i and β_i, all of which are constants depending only on the nodal coordinates (x_i, y_i) for that element as is clear from (4.24). For the constitutive model of homogeneous linear elasticity in 2-D Cartesian coordinates, the elastic constants C_{11}, C_{12}, and C_{66} are independent of position and are of the form in (4.64) for orthotropic or higher symmetry:

$$[\boldsymbol{C}] = \begin{bmatrix} C_{11} & C_{12} & 0 \\ C_{12} & C_{22} & 0 \\ 0 & 0 & C_{66} \end{bmatrix}. \tag{4.90}$$

Using (4.89) and (4.90), the diagonal components of the linear triangular element stiffness matrix are obtained analytically from the expression $k_{ij}^{(e)} = B_{ik}^{\mathrm{T}} C_{kl} B_{lj} A_e$ as follows:

$$\begin{aligned} &4A_e k_{11}^{(e)} = C_{11}\beta_1^2 + C_{66}\gamma_1^2, \quad 4A_e k_{22}^{(e)} = C_{22}\gamma_1^2 + C_{66}\beta_1^2, \quad 4A_e k_{33}^{(e)} = C_{11}\beta_2^2 + C_{66}\gamma_2^2, \\ &4A_e k_{44}^{(e)} = C_{22}\gamma_2^2 + C_{66}\beta_2^2, \quad 4A_e k_{55}^{(e)} = C_{11}\beta_3^2 + C_{66}\gamma_3^2, \quad 4A_e k_{66}^{(e)} = C_{22}\gamma_3^2 + C_{66}\beta_3^2. \end{aligned} \tag{4.91}$$

The same approach produces the following off-diagonal stiffness components:

$$
\begin{aligned}
4A_e k_{12}^{(e)} &= C_{12}\beta_1\gamma_1 + C_{66}\beta_1\gamma_1, & 4A_e k_{13}^{(e)} &= C_{11}\beta_1\beta_2 + C_{66}\gamma_1\gamma_2, \\
4A_e k_{14}^{(e)} &= C_{12}\beta_1\gamma_2 + C_{66}\beta_2\gamma_1, & 4A_e k_{15}^{(e)} &= C_{11}\beta_1\beta_3 + C_{66}\gamma_1\gamma_3, \\
4A_e k_{16}^{(e)} &= C_{12}\beta_1\gamma_3 + C_{66}\beta_3\gamma_1, & 4A_e k_{23}^{(e)} &= C_{12}\beta_2\gamma_1 + C_{66}\beta_1\gamma_2, \\
4A_e k_{24}^{(e)} &= C_{22}\gamma_1\gamma_2 + C_{66}\beta_1\beta_2, & 4A_e k_{25}^{(e)} &= C_{12}\beta_3\gamma_1 + C_{66}\beta_1\gamma_3, \\
4A_e k_{26}^{(e)} &= C_{22}\gamma_1\gamma_3 + C_{66}\beta_1\beta_3, & 4A_e k_{34}^{(e)} &= C_{12}\beta_2\gamma_2 + C_{66}\beta_2\gamma_2, \\
4A_e k_{35}^{(e)} &= C_{11}\beta_2\beta_3 + C_{66}\gamma_2\gamma_3, & 4A_e k_{36}^{(e)} &= C_{12}\beta_2\gamma_3 + C_{66}\beta_3\gamma_2, \\
4A_e k_{45}^{(e)} &= C_{12}\beta_3\gamma_2 + C_{66}\beta_2\gamma_3, & 4A_e k_{46}^{(e)} &= C_{22}\gamma_2\gamma_3 + C_{66}\beta_2\beta_3, \\
4A_e k_{56}^{(e)} &= C_{12}\beta_3\gamma_3 + C_{66}\beta_3\gamma_3.
\end{aligned}
\tag{4.92}
$$

Only the components above and to the right of the diagonal are listed in (4.92) since $\boldsymbol{k}^{(e)}$ is symmetric: $k_{ij}^{(e)} = k_{ji}^{(e)}$. Units of components of $\boldsymbol{k}^{(e)}$ in (4.91) and (4.92) have physical dimensions of force per unit area. The usual stiffness dimensions of force per unit length (e.g., as in a spring constant k) are obtained via multiplication by the body's out-of-plane thickness which is constant over the domain Ω for planar 2-D problems and is typically assigned a value of unity for convenience.

The assembly process for 2-D linear elasticity problems is similar to that for 1-D geometries considered in prior chapters and for 2-D conduction problems considered earlier in Chapter 4. The global system of equations for quasi-static linear elasticity is written as

$$[\boldsymbol{K}]\{\boldsymbol{u}\} = \{\boldsymbol{Q}\} + \{\boldsymbol{f}\} = \{\boldsymbol{F}\}, \tag{4.93}$$

with $\boldsymbol{K}$ the assembled global stiffness matrix, $\boldsymbol{u}$ the column vector of nodal displacements (including both x- and y-components), $\boldsymbol{Q}$ the column vector of contributions of applied tractions resolved at the nodes, and $\boldsymbol{f}$ the column vector of applied body forces resolved at the nodes. The total external force vector is $\boldsymbol{F}$. If n is the total number of nodes in the domain Ω, then each vector in (4.93) is of length $2n$, and the global stiffness matrix is of dimensions $2n \times 2n$. At a given node, contributions from all elements sharing that node sum to produce load vectors $\boldsymbol{Q}$ and $\boldsymbol{f}$. Element-level contributions are defined by the integrals in (4.86) that require integration of the shape functions N_i.

Global stiffness matrix $\boldsymbol{K}$ can be assembled by consideration of element equilibrium equations and mesh connectivity. The assembly process is written symbolically, as usual, as

$$[\boldsymbol{K}] = \bigwedge_{e \in \Omega} [\boldsymbol{k}^{(e)}]. \tag{4.94}$$

As is the case for 2-D heat transfer, in elasticity problems the operation in (4.94) is arduous for a domain meshed by any significant number of elements. Fortunately, this task is performed for the analyst by any modern FEM software.

The remainder of the analysis of a 2-D static linear elasticity problem using FEA follows the same general procedure as that for truss elements and beam elements in respective Chapters 2 and 3. In particular, the key remaining steps in an analysis of a problem of the type shown in Fig. 4.12 are

- **Boundary Conditions**: manipulate the global equations to account for essential boundary conditions on primary solution variables u and υ, moving contributions of nonzero imposed generalized displacements u_i on $\partial\Omega_u$ to the right side of the global system of equations in (4.93);
- **Solve**: solve the condensed global system $[\boldsymbol{K}]\{\boldsymbol{u}\} = \{\boldsymbol{F}\}$ where boundary conditions have been imposed, such that condensed global stiffness matrix $[\boldsymbol{K}]$ is non-singular and invertible;
- **Post-process**: compute secondary nodal and elemental quantities by substituting primary solution variables $(\boldsymbol{u})$ into element equations.

Finally, a few remarks on post-processing are in order for 2-D elasticity problems. Consider determination of the stress $\boldsymbol{\sigma}^{(e)}$ for an element. Noting that Hooke's law $\boldsymbol{\sigma} = \boldsymbol{C\varepsilon}$ of (4.59) holds within any element, the vector of stress compopments σ_{xx}, σ_{yy}, and σ_{xy} for that element can be computed from the primary solution field $\boldsymbol{u}^{(e)}$ referred to local nodes of that element using (4.83) as

$$\{\boldsymbol{\sigma}^{(e)}\} = [\boldsymbol{C}]\{\boldsymbol{\varepsilon}^{(e)}\} = [\boldsymbol{C}][\boldsymbol{B}(x,y)]\{\boldsymbol{u}^{(e)}\}. \tag{4.95}$$

Elasticity matrix $\boldsymbol{C}$ and shape function derivative matrix $\boldsymbol{B}$ also depend on the element number. For linear triangular elements, since the shape functions are linear in x and y, displacement gradients, and therefore strains, are constant within each element. Thus, for heterogeneous displacement profiles, stress components will usually be discontinuous along edges between shared elements when calculated via (4.95). Therefore, nodal strain and stress values are often averaged from neighboring element values as a post-processing option in FEM software.

4.10 ANSYS Example: Planar Elastic Body with Corners

The example described next highlights some concepts from static 2-D elasticity theory using the ANSYS software (v.17.2, 2016). Stress concentrations, mesh refinement, and smoothing of sharp corners in the geometry are demonstrated.

4.10.1 Problem Statement

The dimensions and properties of a planar body are shown in Fig. 4.13. The body is presumably located in the xy-plane and is thin in the out-of-plane (z) direction, implying a plane stress analysis. The body is fully constrained along its left and bottom edges, with displacements $u_x = u_y = 0$. Two point forces are applied in the negative y-direction on the right side, each of magnitude $P/2 = 5$ kN. The isotropic elastic properties are modulus $E = 100$ GPa and Poisson's ratio $\nu = 0.27$. The objectives of this example are as follows. Firstly, we seek the primary solution (i.e., nodal displacements) and maximum effective stress for a coarse mesh. Secondly, we intend

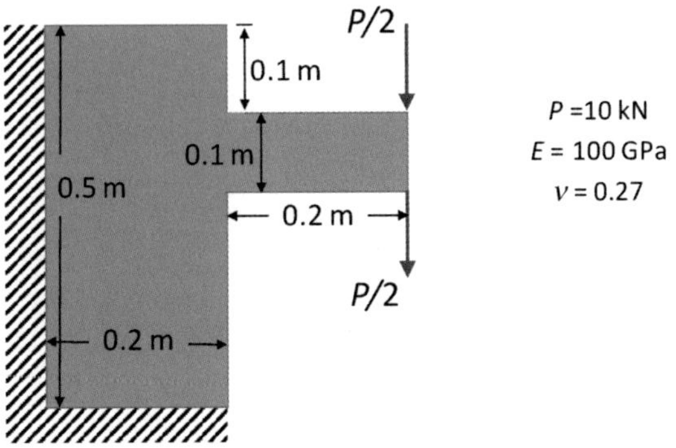

Fig. 4.13 Planar elasticity example problem: geometry, boundary conditions, and properties

to determine of the same solution variables for several progressively finer meshes. Finally, we will use fillets to alleviate stress concentrations at corners.

4.10.2 Pre-processing

Pre-processing involves choice of element type and material properties, prescription of plane stress conditions, mesh creation, and imposition of boundary conditions. The FE analysis in the ANSYS package here commences via selection of an element type for planar continuum elasticity, the `quad 4 node 182` element:

```
Main Menu > Preprocessor > Element Type > Add/Edit/Delete
> Add > Quad 4 node 182
```

The choice is demonstrated in Fig. 4.14 which also shows verification of the element option for plane stress. The material model is isotropic linear elasticity with properties enacted by

```
Main Menu > Preprocessor > Material Properties > Material
Models > Structural > Linear > Elastic > Isotropic
```

The elastic modulus E and Poisson's ratio ν are entered as 100e9 Pa and 0.27 in accordance with Fig. 4.13. The planar body is created by generating two rectangles then adding them together. Rectangles are produced via two iterations of the sequence

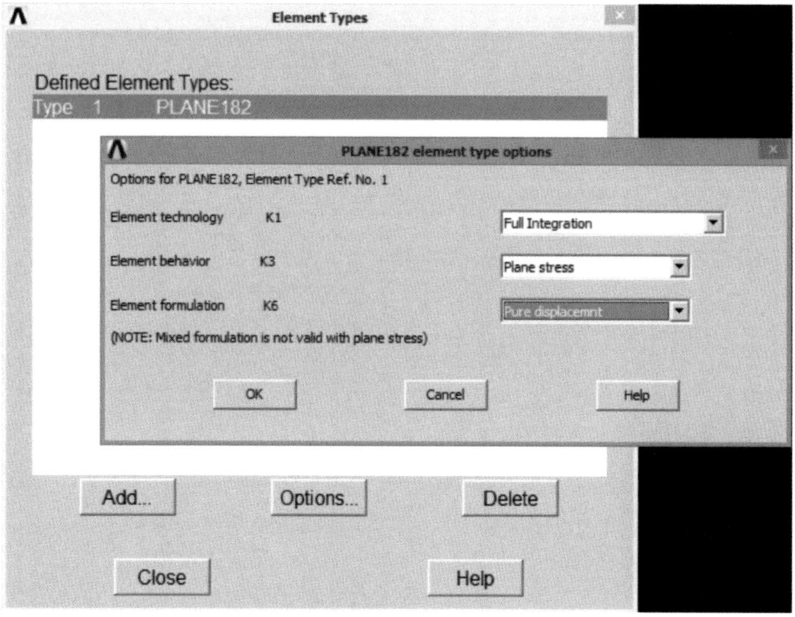

Fig. 4.14 Planar elasticity example problem: element data

```
Main Menu > Preprocessor > Modeling > Create > Areas >
Rectangle > By Dimensions
```

Refer to Fig. 4.15. Addition of the two areas to produce a single body is executed by

```
Main Menu > Preprocessor > Modeling > Operate > Booleans >
Add > Areas
```

Pick both rectangles then click "OK". The next step is choice of the static analysis type:

```
Main Menu > Solution > Analysis Type > New Analysis >
Static
```

For this problem that will involve remeshing, it is more efficient to apply boundary conditions to the underlying geometry than to nodes of the FE mesh. The former method enables maintenance of the same boundary conditions upon remeshing, whereas the latter would require us to update the boundary conditions each time the

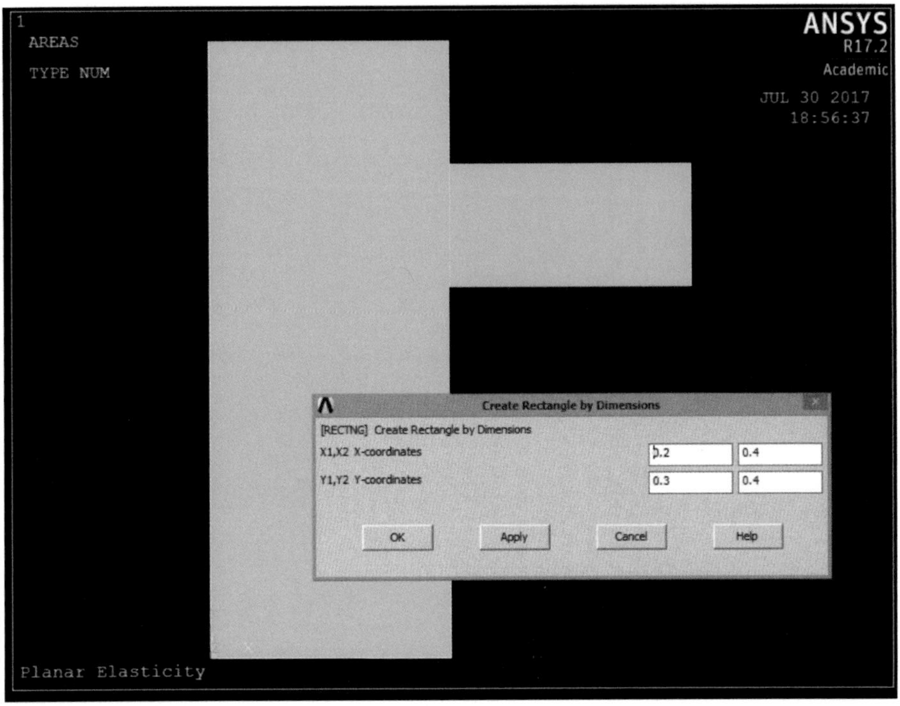

Fig. 4.15 Planar elasticity example problem: geometric modeling

mesh is changed. Referring to Fig. 4.13, first impose the essential boundary conditions of null displacements on the left and bottom edges of the body:

```
Main Menu > Preprocessor > Define Loads > Apply >
Structural > Displacement > On Lines
```

Pick the two appropriate lines and constrain all degrees of freedom (DOF) to be zero. The natural BCs corresponding to the point forces $P/2$ on the right side are then prescribed:

```
Main Menu > Preprocessor > Define Loads > Apply >
Structural > Force/Moment > On Keypoints
```

Enter the FY values of $P/2 = -5$ kN for each of the corresponding keypoints. The mesh size is set as follows:

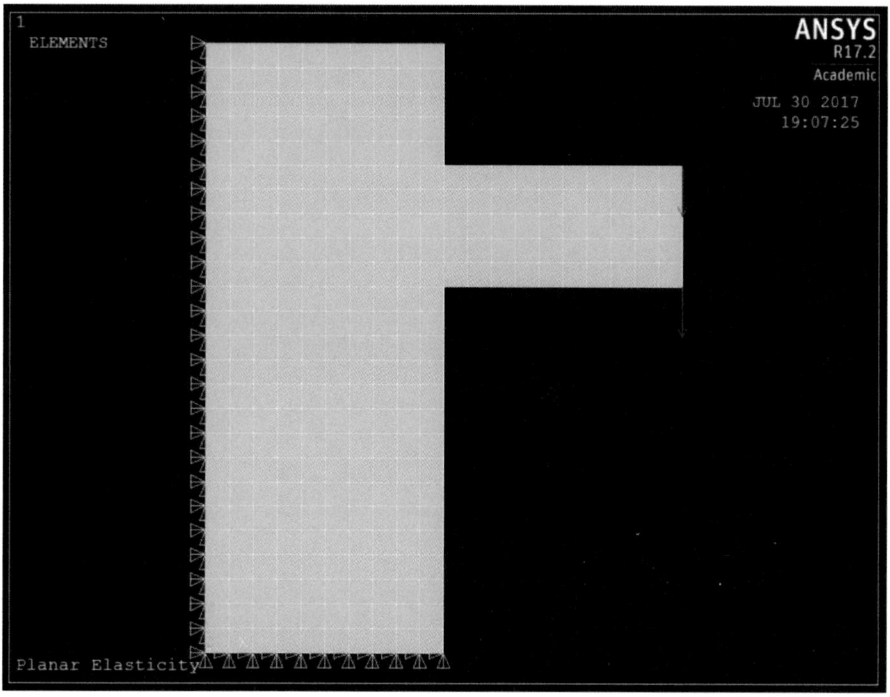

Fig. 4.16 Planar elasticity example problem: mesh and boundary conditions

```
Main Menu > Preprocessor > Meshing > Size Cntrls > Manual
Size > Areas > All Areas
```

A value of 0.02 is entered for the element edge length that will produce a rather coarse initial mesh. The (free) mesh is then generated via

```
Main Menu > Preprocessor > Mesh > Areas > Free > Pick
```

Click "OK" to produce the mesh shown in Fig. 4.16, which also indicates the boundary conditions.

4.10.3 Solutions, Modifications, and Post-processing

The FE solution for the initial mesh and geometry is obtained by

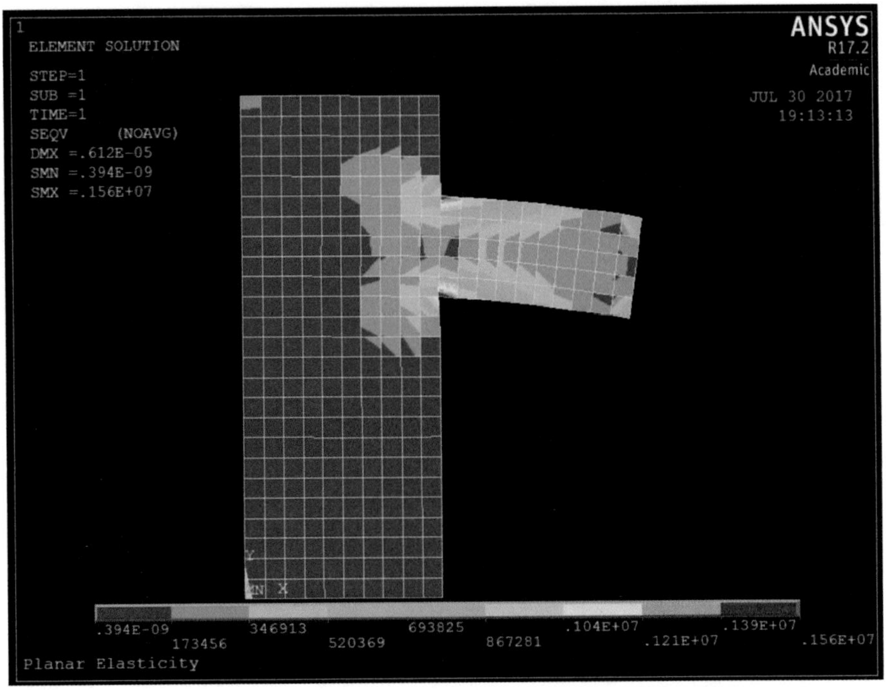

Fig. 4.17 Planar elasticity example problem: stress contour, original mesh

```
Main Menu > Solution > Solve > Current LS > OK
```

For post-processing, we seek the deformed shape of the body, the maximum deflection over all nodes in the body, and the distribution of stress in the body. First, the deformed structure can be viewed in superposition with the original body via

```
Main Menu > General Postproc > Plot Results > Deformed
Shape
```

Choose "Deformed and Undeformed Edge". The displacements are magnified by the ANSYS GUI to enable visualization; otherwise, the original and deformed geometries would be too similar for us to observe any differences. A contour plot of von Mises stress in the elements is created by the sequence below:

```
Main Menu > General Postproc > Plot Results > Element
Solution > von Mises stress
```

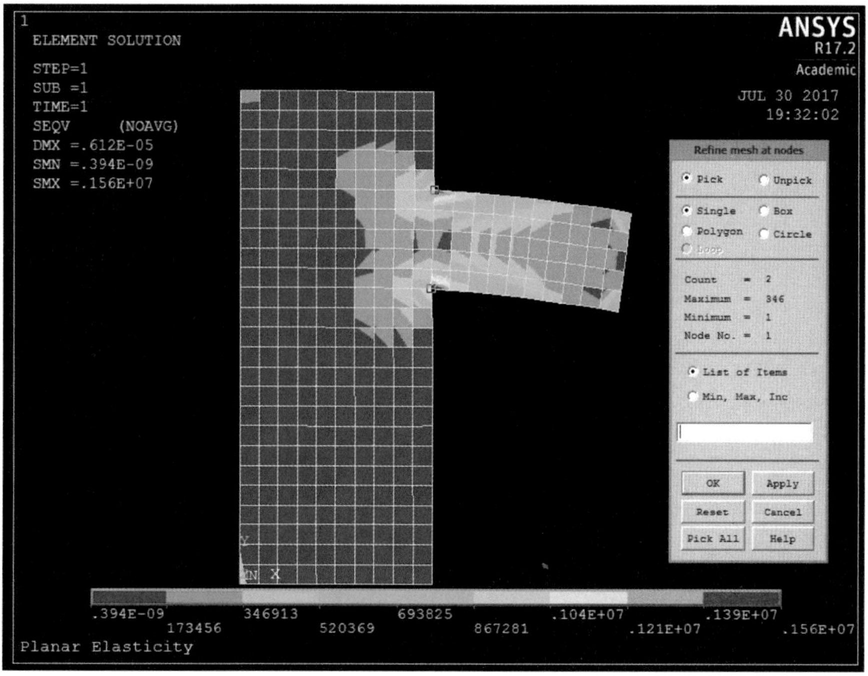

Fig. 4.18 Planar elasticity example problem: mesh refinement procedure

The von Mises stress is a non-negative scalar measure of the intensity of the stress tensor, excluding effects of hydrostatic pressure. In many engineering materials such as ductile metals, plastic yielding tends to occur when the von Mises stress exceeds a threshold value. For a plane stress problem, the von Mises stress is computed from components of the stress tensor as

$$\sigma_v = \left[(\sigma_{xx})^2 + (\sigma_{yy})^2 - \sigma_{xx}\sigma_{yy} + 3(\sigma_{xy})^2\right]^{1/2}. \tag{4.96}$$

The maximum value of the von Mises stress over all elements (or Gauss points) in the mesh is defined as

$$\sigma_m = \max_{(e)} \sigma_v^{(e)}. \tag{4.97}$$

From Fig. 4.17, the maximum magnitude of displacement is 6.12 micrometers, and the maximum von Mises stress is $\sigma_m = 1.56$ MPa. Stress concentrations are evident at the corners where the two original rectangles intersect, i.e., at the top and bottom of the cantilevered section's left boundary. The mesh is refined by adding more elements near these points of concentrated stress by the following command sequence:

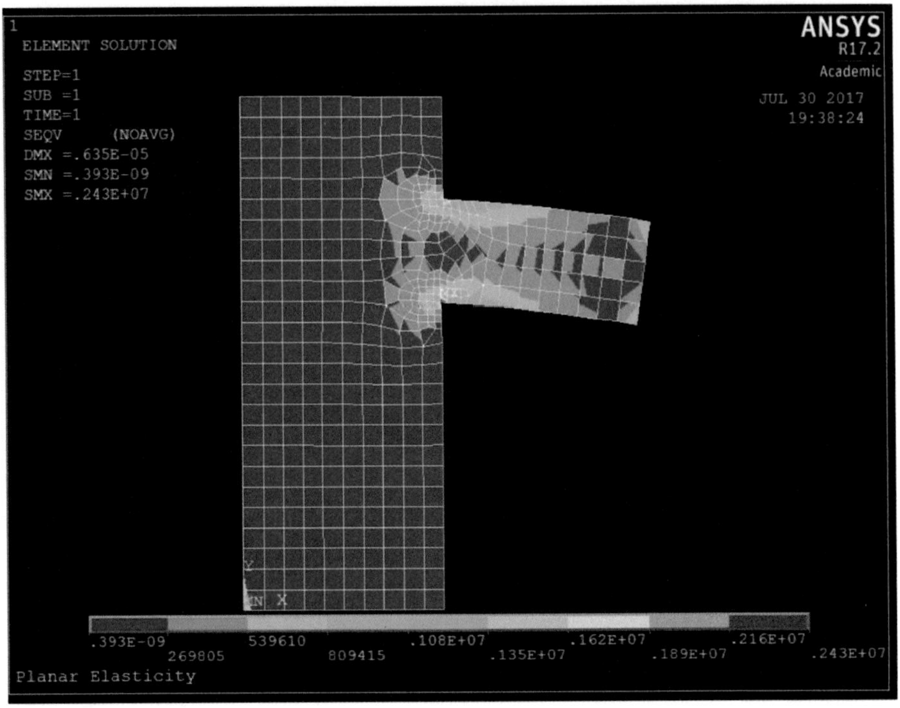

Fig. 4.19 Planar elasticity example problem: stress contour, first refinement

```
Main Menu > Preprocessor > Meshing > Modify Mesh > Refine
At > Nodes
```

Nodes at the corner points are chosen as in Fig. 4.18, and the default level of "1 (Minimal)" is prescribed for refinement. The solution steps are then repeated with this refined mesh, noting that no updates to boundary conditions are necessary. The end result is the contour plot of von Mises stress in Fig. 4.19. The maximum displacement is 6.35 micrometers, which is not too different from the primary solution with the original coarse mesh, but the maximum stress is now 2.43 MPa, much larger than the original result of 1.56 MPa. Stress concentrations occur at the same corner points as in the first analysis.

Two more iterations of refinement are executed, the first locally at the same two corner points, while the second is prescribed as a global refinement. The maximum effective stress is recorded after each iteration, as is the number of elements in the mesh. The contour of stress for the final/third refinement is given in Fig. 4.20. The maximum von Mises stress is $\sigma_m = 5.30$ MPa for a mesh consisting of 2040 elements. The corresponding maximum displacement is 6.59 micrometers. Clearly, the

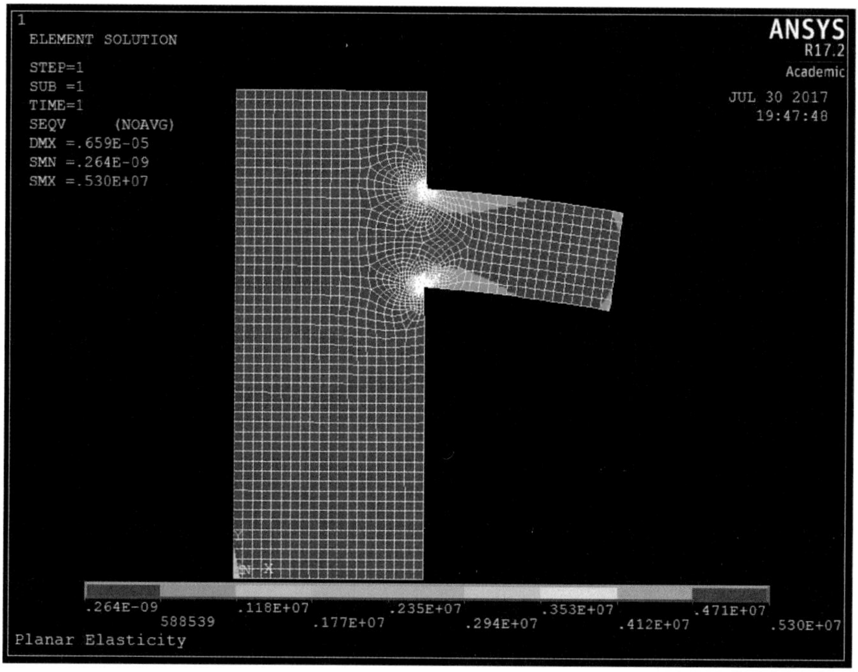

Fig. 4.20 Planar elasticity example problem: stress contour, third refinement

solution for the stress field is not converging with decreasing mesh size, i.e., with increasing refinement. Rather, the elasticity solution for this problem suffers from stress singularities at the corner points of stress concentration where infinite stress magnitudes are mathematical artifacts for this geometry and loading protocol.

Stress concentrations can be reduced or eliminated by careful use of fillets that smooth out sharp corners in the geometry. Next in this example, we add fillets at the two corners where maximum stresses arise. First, the original mesh is removed from the geometry:

```
Main Menu > Preprocessor > Meshing > Clear > Areas
```

Pick the lone existing area and click "OK" to delete the mesh. The geometry and boundary conditions remain intact. A fillet is created at a corner by the command sequence

```
Main Menu > Preprocessor > Modeling > Create > Lines >
Line Fillet
```

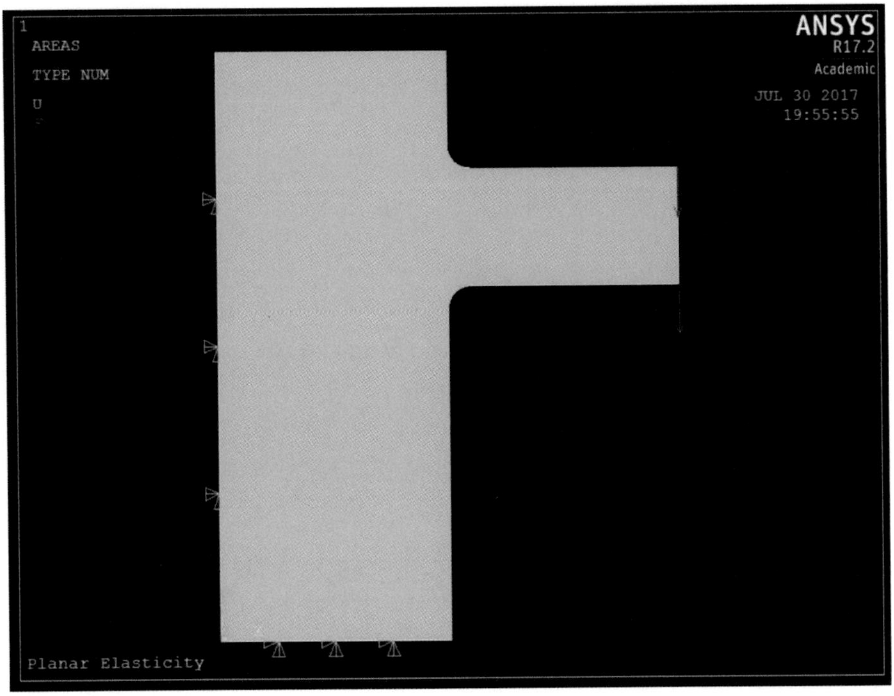

Fig. 4.21 Planar elasticity example problem: geometry and boundary conditions with fillets

Pick the two lines intersecting the corner point and enter a fillet radius RAD of 0.02. The smaller the radius, the sharper the rounded corner. Then create an area enclosed by the fillet region as follows:

```
Main Menu > Preprocessor > Modeling > Create > Areas >
Arbitrary > By Lines
```

The three lines (two straight plus the curved line at the fillet) are then selected. The above process is repeated at the second corner of concentrated stress. Finally, the two new areas, one for each fillet, are added to the area associated with the original body:

```
Main Menu > Preprocessor > Modeling > Operate > Booleans >
Add > Areas
```

Pick all three areas and then click "OK". Visualization of the resulting smoothed body in Fig. 4.21 verifies that the original boundary conditions are still in place.

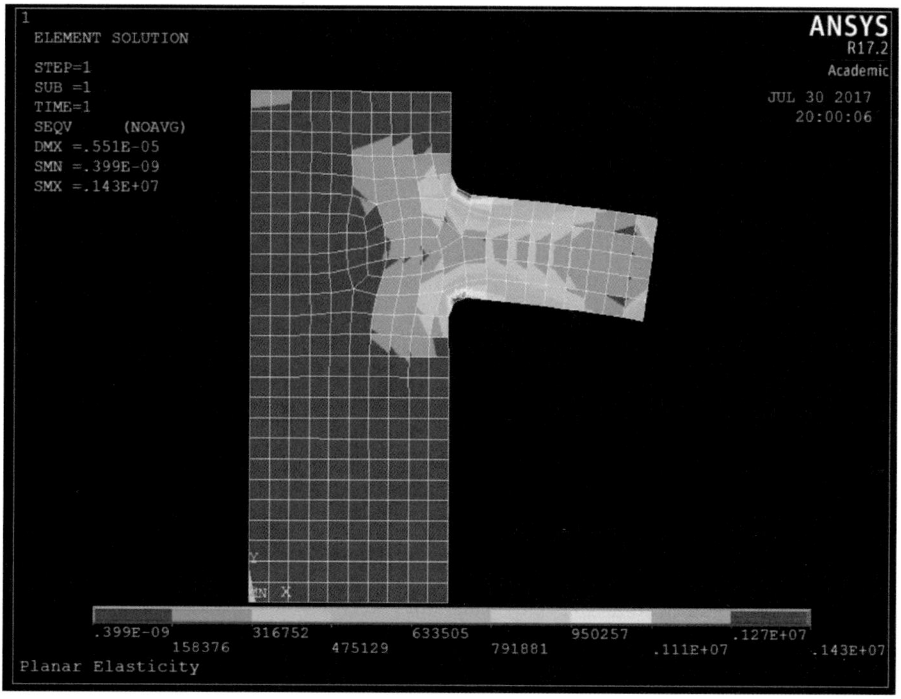

Fig. 4.22 Planar elasticity example problem: stress contour, mesh with fillets

Next, the body with fillets is meshed by the same procedure followed for the original body. The element size is set to 0.02:

```
Main Menu > Preprocessor > Meshing > Size Cntrls > Manual
Size > Areas > All Areas
```

The first mesh for the smoothed domain is created by

```
Main Menu > Preprocessor > Mesh > Areas > Free > Pick
```

A solution for the current load step is obtained following the same procedure used for the original mesh without fillets. The resulting contour of von Mises stress for the body with fillets is shown in Fig. 4.22. The maximum deflection is 5.51 micrometers, and the maximum von Mises stress is $\sigma_m = 1.43$ MPa.

The same sequence of progressive mesh refinement and post-processing is enacted for the body with fillets. Three iterations are performed; a contour plot of stress for the final iteration, with 2052 elements, is shown in Fig. 4.23. The max-

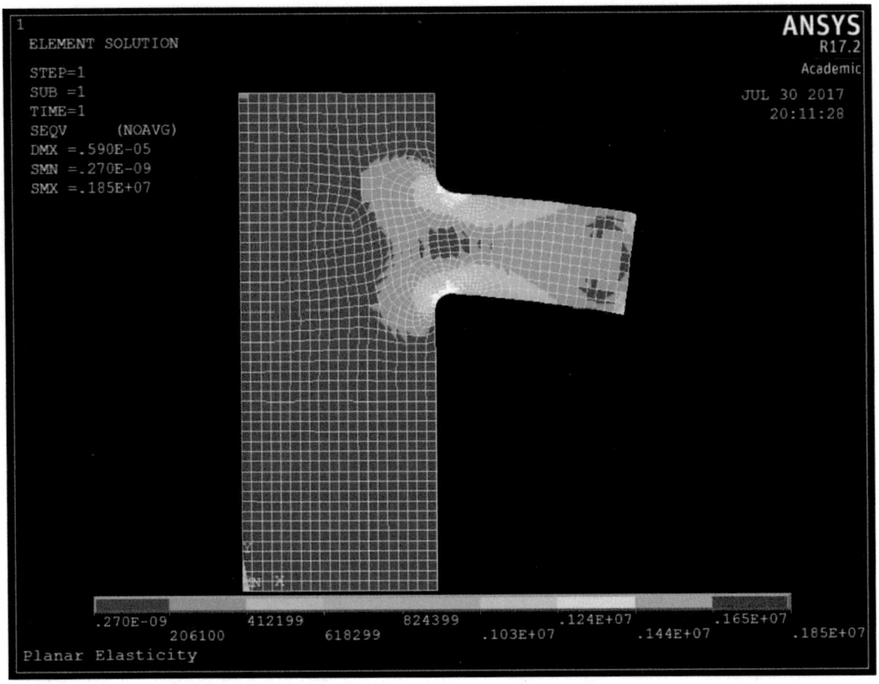

Fig. 4.23 Planar elasticity example problem: stress contour, third refinement with fillets

imum deflection in this case is 5.90 micrometers, the maximum stress 1.85 MPa. Comparison with Fig. 4.20 demonstrates the reduced stress concentration for the model with fillets relative to the model with sharp corners.

Shown in Fig. 4.24 is the maximum stress versus number of elements for the two bodies, one without fillets and one with fillets. The model without fillets demonstrates continuously, and significantly, increasing stress with increasing mesh density. On the other hand, the model with fillets exhibits only slight increases in stress with increasing number of elements, suggesting convergence to an accurate result.

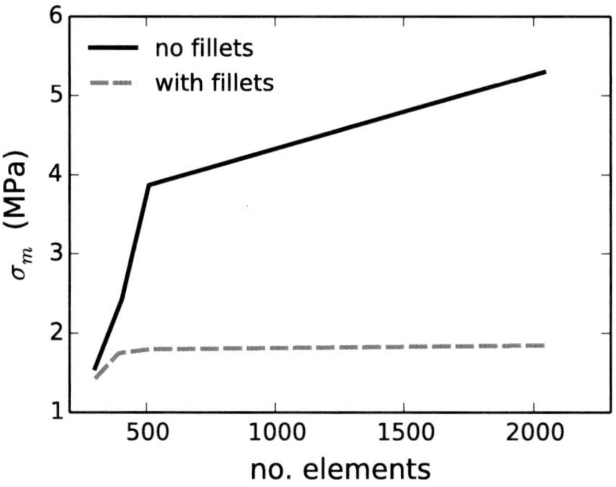

Fig. 4.24 Planar elasticity example problem: maximum effective stress versus mesh size

4.11 Problems

4.11.1. Consider triangular finite elements with linear shape functions.
a. Show that for an element with three nodes at coordinates (x_i, y_i) and edges enclosing area A, the shape functions are of the following form for $i = 1, 2, 3$:

$$N_i(x,y) = \frac{1}{2A}(\alpha_i + \beta_i x + \gamma_i y),$$

$$\begin{bmatrix} \alpha_1 & \alpha_2 & \alpha_3 \\ \beta_1 & \beta_2 & \beta_3 \\ \gamma_1 & \gamma_2 & \gamma_3 \end{bmatrix} = \begin{bmatrix} (x_2y_3 - x_3y_2) & (x_3y_1 - x_1y_3) & (x_1y_2 - x_2y_1) \\ y_2 - y_3 & y_3 - y_1 & y_1 - y_2 \\ x_3 - x_2 & x_1 - x_3 & x_2 - x_1 \end{bmatrix}.$$

b. For a triangle with (x_i, y_i) corner points [(0,0),(3,0),(2,2)], evaluate all coefficients for all three shape functions and demonstrate that the shape functions $N_i(x_j, y_j)$ have the correct local properties. [Hint: consider the Kronecker delta properties in equation (4.19).]

4.11.2. Consider the 2-D domain Ω shown in Fig. 4.25 with prescribed nonzero (isotropic) conductivity components $k = k_{xx} = k_{yy}$ and essential or natural boundary conditions imposed on various sides.
a. Using finite element software and 2-D thermal (heat conduction) elements, solve for the steady state temperature field $T(x, y)$. Obtain a contour plot of the temperature distribution. [Hint: be careful regarding the sign convention used for heat flux boundary conditions in the FE software.]
b. Repeat the analysis from part **a.** using a finer mesh. Supply another contour plot of the temperature field. State if your solution appears sensitive to mesh density or does not, and thus whether or not your mesh is sufficiently fine.

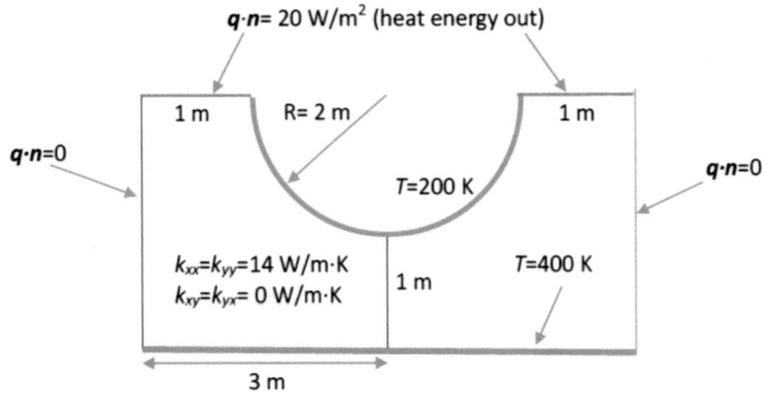

Fig. 4.25 Problem 4.11.2

c. Repeat the analysis from part **a.** using a different value of $k = k_{xx} = k_{yy}$. Does the solution vary significantly, or at all, with variations in k? Why or why not?

References

1. L.E. Malvern, *Introduction to the Mechanics of a Continuous Medium* (Prentice-Hall, Englewood Cliffs, NJ, 1969)
2. J.D. Clayton, *Differential Geometry and Kinematics of Continua* (World Scientific, Singapore, 2014)
3. J.D. Clayton, *Nonlinear Mechanics of Crystals* (Springer, Dordrecht, 2011)
4. C. Teodosiu, *Elastic Models of Crystal Defects* (Springer, Berlin, 1982)

Chapter 5
Axisymmetric Analysis

Abstract Axisymmetric problems are addressed in this chapter, wherein generally, field variables depend on two coordinates: the radial position r and the axial position z. These problems are a simpler case of fully 3-D cylindrical geometry, where for the axisymmetric case dependence on angular coordinate θ is absent or suppressed. First, cylindrical coordinates and corresponding differential operators are presented. Next, governing equations for steady-state heat transfer are derived in axisymmetric domains. Strong form equations, weak form equations, and interpolation functions are examined. Finite element equations of thermal equilibrium are obtained in vector-matrix form for axisymmetric problems. Similar developments are then presented for axisymmetric problems in linear elasticity: governing differential equations, strong and weak forms, and finite element equations for static equilibrium are discussed in sequence. A few potentially important issues emerging in axisymmetric problems that do not arise in planar 2-D problems are noted.

This chapter begins with coverage of essential background mathematics on cylindrical coordinate systems. Axisymmetric problems for steady heat conduction and quasi-static linear elasticity are then addressed in turn.

5.1 Coordinate Systems

Axisymmetric problems invoke cylindrical coordinates with the restriction that field variables do not depend on the angular coordinate θ. This restriction corresponds to existence of a primary axis of rotational symmetry that serves an axis of transverse isotropy for material properties. In this book, the axis of rotational symmetry is taken as the z-axis. Cylidrical coordinates (r, θ, z) are related to Cartesian 3-D coordinates (x, y, z) as follows:

$$x = r\cos\theta, \qquad y = r\sin\theta, \qquad z = z; \tag{5.1}$$

$$r = \sqrt{x^2 + y^2} \geq 0, \qquad \theta = \tan^{-1}(y/x) \in (-\pi, \pi]. \tag{5.2}$$

Vectors and tensors will be described in so-called physical components, whereby basis vectors are all dimensionless, orthogonal, and of unit length. Specifically, let $\boldsymbol{e}_r$, $\boldsymbol{e}_\theta$, and $\boldsymbol{e}_z$ denote the triad of mutually perpendicular unit basis vectors in the radial, circumferential, and axial directions, respectively. Then the gradient operator in cylindrical coordinates is [1, 2]

$$\nabla(\cdot) = \frac{\partial(\cdot)}{\partial r}\boldsymbol{e}_r + \frac{1}{r}\frac{\partial(\cdot)}{\partial\theta}\boldsymbol{e}_\theta + \frac{\partial(\cdot)}{\partial z}\boldsymbol{e}_z. \tag{5.3}$$

The gradient of a scalar field $f(r,\theta,z)$ is then the vector field

$$\nabla f = \frac{\partial f}{\partial r}\boldsymbol{e}_r + \frac{1}{r}\frac{\partial f}{\partial\theta}\boldsymbol{e}_\theta + \frac{\partial f}{\partial z}\boldsymbol{e}_z. \tag{5.4}$$

The Laplacian of a scalar field $f(r,\theta,z)$ is the scalar field [1, 2]

$$\nabla^2 f = \frac{1}{r}\left(r\frac{\partial f}{\partial r}\right) + \frac{1}{r^2}\frac{\partial^2 f}{\partial\theta^2} + \frac{\partial^2 f}{\partial z^2}. \tag{5.5}$$

The divergence of a vector field $\boldsymbol{v}(r,\theta,z)$ with physical components v_r, v_θ, and v_z is the scalar field [1, 2]

$$\nabla\cdot\boldsymbol{v} = \frac{1}{r}\frac{\partial}{\partial r}(rv_r) + \frac{1}{r}\frac{\partial v_\theta}{\partial\theta} + \frac{\partial v_z}{\partial z}. \tag{5.6}$$

The divergence of a second-order tensor field $\boldsymbol{T}(r,\theta,z)$ with physical components T_{rr}, $T_{r\theta}$, etc., is the vector field [1, 2]

$$\begin{aligned}\nabla\cdot\boldsymbol{T} = &\left[\frac{1}{r}\frac{\partial}{\partial r}(rT_{rr}) + \frac{1}{r}\frac{\partial T_{\theta r}}{\partial\theta} + \frac{\partial T_{zr}}{\partial z} - \frac{T_{\theta\theta}}{r}\right]\boldsymbol{e}_r \\ &+ \left[\frac{1}{r}\frac{\partial T_{\theta\theta}}{\partial\theta} + \frac{\partial T_{z\theta}}{\partial z} + \frac{1}{r}\frac{\partial}{\partial r}(rT_{r\theta}) + \frac{T_{\theta r}}{r}\right]\boldsymbol{e}_\theta \\ &+ \left[\frac{\partial T_{zz}}{\partial z} + \frac{1}{r}\frac{\partial}{\partial r}(rT_{rz}) + \frac{1}{r}\frac{\partial T_{\theta z}}{\partial\theta}\right]\boldsymbol{e}_z.\end{aligned} \tag{5.7}$$

In axisymmetric problems, $f = f(r,z)$, $\boldsymbol{v} = \boldsymbol{v}(r,z)$, $\boldsymbol{T} = \boldsymbol{T}(r,z)$, and operations in (5.3)–(5.7) simplify, respectively, as follows:

$$\nabla(\cdot) = \frac{\partial(\cdot)}{\partial r}\boldsymbol{e}_r + \frac{\partial(\cdot)}{\partial z}\boldsymbol{e}_z; \tag{5.8}$$

$$\nabla f = \frac{\partial f}{\partial r}\boldsymbol{e}_r + \frac{\partial f}{\partial z}\boldsymbol{e}_z; \tag{5.9}$$

$$\nabla^2 f = \frac{1}{r}\left(r\frac{\partial f}{\partial r}\right) + \frac{\partial^2 f}{\partial z^2}; \tag{5.10}$$

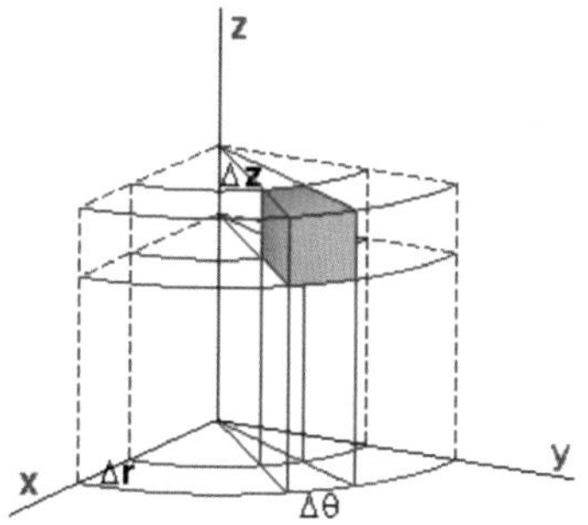

Fig. 5.1 Cylindrical coordinates (r,θ,z) and Cartesian coordinates (x,y,z), with volume element (shaded) $\Delta V = r\cdot\Delta r\cdot\Delta\theta\cdot\Delta z$ [image adapted from `www.math.byu.edu`]

$$\nabla\cdot\boldsymbol{v} = \frac{1}{r}\frac{\partial}{\partial r}(rv_r) + \frac{\partial v_z}{\partial z}; \tag{5.11}$$

$$\begin{aligned}\nabla\cdot\boldsymbol{T} = &\left[\frac{1}{r}\frac{\partial}{\partial r}(rT_{rr}) + \frac{\partial T_{zr}}{\partial z} - \frac{T_{\theta\theta}}{r}\right]\boldsymbol{e}_r \\ &+ \left[\frac{\partial T_{z\theta}}{\partial z} + \frac{1}{r}\frac{\partial}{\partial r}(rT_{r\theta}) + \frac{T_{\theta r}}{r}\right]\boldsymbol{e}_\theta + \left[\frac{\partial T_{zz}}{\partial z} + \frac{1}{r}\frac{\partial}{\partial r}(rT_{rz})\right]\boldsymbol{e}_z.\end{aligned} \tag{5.12}$$

Notice that although no components of $\boldsymbol{v}$ or $\boldsymbol{T}$ can vary with angular coordinate θ, individual components with θ indices (e.g., $v_\theta, T_{r\theta}, T_{\theta\theta},\ldots$) do not vanish, in general for axisymmetric problems. A differential volume element, for example as shown in Fig. 5.1, is, in 3-D cylindrical coordinates,

$$\mathrm{d}\Omega = r\,\mathrm{d}r\,\mathrm{d}\theta\,\mathrm{d}z. \tag{5.13}$$

The same volume element applies for axisymmetric bodies.

5.2 Heat Conduction: Governing Equations

Consider an axisymmetric domain Ω with boundary $\partial\Omega$, where cylindrical coordinates r, θ, and z cover the domain. The unit outward normal vector to Ω is $\boldsymbol{n}(r,z)$. Denote the temperature field by $T(r,z)$, the heat flux vector by $\boldsymbol{q}(r,z)$, and the scalar heat source by $f(r,z)$.

Analogously to planar 2-D problems considered in §4.1, boundary conditions on $\partial\Omega$ are of two possible types:

- **Essential**: temperature prescribed on $\partial\Omega_T$;
- **Natural**: scalar heat flux prescribed on $\partial\Omega_q$.

The scalar heat flux on the boundary is again defined as $q_n = \boldsymbol{q}\cdot\boldsymbol{n}$.

The local balance of energy for static heat conduction, i.e., thermal equilibrium, is written in vector form identically to (4.1) [1, 3]:

$$\nabla \cdot \boldsymbol{q} = f, \tag{5.14}$$

where for fully 3-D domains, $\nabla(\cdot)$ is the gradient operator in cylindrical coordinates of (5.3) and "$\cdot$" the scalar product of vectors, such that $\nabla \cdot (\cdot)$ is the divergence of a cylindrical vector field as in (5.6). In axisymmetric problems, reductions in (5.8) and (5.11) hold, and this equation is

$$\frac{1}{r}\frac{\partial}{\partial r}[rq_r(r,z)] + \frac{\partial q_z(r,z)}{\partial z} = f(r,z). \tag{5.15}$$

Recall from Chapter 4 that heat flux $\boldsymbol{q}$ is positive in the direction of energy flow, such that boundary surface flux q_n is positive when energy leaves the body. Heat source f is positive in sign if it would tend to increase the internal energy of the body.

The constitutive model considered herein is, as in Chapter 4, linear Fourier heat conduction. Denote by $\boldsymbol{k}(r,z)$ the symmetric thermal conductivity matrix, of size 2×2 for axisymmetric problems where the domain is transversely isotropic with respect to material properties:

$$[\boldsymbol{k}] = \begin{bmatrix} k_{11} & 0 \\ 0 & k_{22} \end{bmatrix} = \begin{bmatrix} k_{rr} & 0 \\ 0 & k_{zz} \end{bmatrix}. \tag{5.16}$$

Conductivity in radial and axial directions is measured by k_{rr} and k_{zz}, respectively. When $\boldsymbol{k} = k\mathbf{I}$ (i.e., $k_{ij} = k\delta_{ij}$) with $k(r,z)$ a scalar, heat conduction is isotropic. Fourier's law is the following constitutive equation in vector-matrix form, similar to (4.4):

$$\{\boldsymbol{q}\} = -[\boldsymbol{k}]\{\nabla T\} \leftrightarrow \begin{Bmatrix} q_r \\ q_z \end{Bmatrix} = -\begin{bmatrix} k_{rr} & 0 \\ 0 & k_{zz} \end{bmatrix} \begin{Bmatrix} \partial T/\partial r \\ \partial T/\partial z \end{Bmatrix}. \tag{5.17}$$

The gradient operation in (5.8) for axisymmetric domains has been invoked for $\nabla T(r,z)$. Recall that since components of $\boldsymbol{k}$ are usually positive in sign, the positive heat flux is most often directed from hot to cold locations of the body. Dimensions of heat flux, heat source, and thermal conductivity are respectively energy rate per unit area, energy rate per unit volume, and energy rate per unit length per unit temperature.

The strong form of the problem statement for steady state conduction is obtained by substituting (5.17) into (5.14):

$$-\nabla \cdot (\boldsymbol{k}\nabla T) = f, \tag{5.18}$$

or in scalar form for axisymmetric problems:

$$-\frac{1}{r}\frac{\partial}{\partial r}\left(rk_{rr}\frac{\partial T}{\partial r}\right) - \frac{\partial}{\partial z}\left(k_{zz}\frac{\partial T}{\partial z}\right) = f. \tag{5.19}$$

When the conductivity is both constant and isotropic, i.e., $\boldsymbol{k} = k\mathbf{I}$ with k =constant, (5.19) reduces to the Poisson equation in cylindrical coordinates without θ dependence:

$$\frac{\partial^2 T}{\partial r^2} + \frac{1}{r}\frac{\partial T}{\partial r} + \frac{\partial^2 T}{\partial z^2} = -\frac{f}{k} \leftrightarrow \nabla^2 T = -f/k. \tag{5.20}$$

The Laplacian operator is $\nabla^2(\cdot)$ and here obeys (5.10). Furthermore, (5.20) degenerates to the Laplace equation when scalar point sources vanish, i.e., when the function $f = 0$:

$$\frac{\partial^2 T}{\partial r^2} + \frac{1}{r}\frac{\partial T}{\partial r} + \frac{\partial^2 T}{\partial z^2} = 0 \leftrightarrow \nabla^2 T = 0. \tag{5.21}$$

In this particular simplest case in (5.21), the steady state solution does not depend on the value of conductivity k.

Most of the treatment in this chapter will address the more general case in (5.19) with nonzero f and anisotropic conductivity. In this more general case, the complete strong form for the problem statement seeks the temperature field $T(r,z)$ in domain Ω that satisfies second-order partial differential equation (5.19), given boundary conditions on $\partial\Omega$, imposed heat source field f, and conductivity matrix $\boldsymbol{k}$ that may generally depend on coordinates r and z.

5.3 Heat Conduction: Weak Form

Prior to derivation of the weak form equation, we again review several identities from vector calculus needed later, here with particular attention given to cylindrical domains with an axis of symmetry. Consider a compact domain Ω with unit outward normal $\boldsymbol{n}$ on boundary $\partial\Omega$, and let $\boldsymbol{v}$ be a continuously differentiable vector field in this domain. The generalized theorem of Gauss is identical to (4.9) when written in coordinate-free vector form, regardless of choice of (cylindrical) coordinate system [1, 2]:

$$\int_\Omega \nabla \cdot \boldsymbol{v}\, \mathrm{d}\Omega = \oint_{\partial\Omega} \boldsymbol{v} \cdot \boldsymbol{n}\, \mathrm{d}\partial\Omega, \tag{5.22}$$

where $\oint(\cdot)$ denotes integration over the boundary enclosing Ω. In cylindrical coordinates, let volume Ω be covered by coordinates (r,θ,z), and let $\mathrm{d}S$ denote a differential area element on $\partial\Omega$ enclosing Ω. Then in the context of fully 3-D cylindrical geometries, (5.22) becomes

$$\int_\Omega \left[\frac{1}{r}\frac{\partial(rv_r)}{\partial r} + \frac{\partial v_\theta}{\partial \theta} + \frac{\partial v_z}{\partial z}\right] r\mathrm{d}r\mathrm{d}\theta\mathrm{d}z = \oint_{\partial\Omega} (v_r n_r + v_\theta n_\theta + v_z n_z)\mathrm{d}S. \tag{5.23}$$

For axisymmetric problems, (5.22) and (5.23) reduce to

$$2\pi \int_\Omega \left[\frac{1}{r}\frac{\partial(rv_r)}{\partial r} + \frac{\partial v_z}{\partial z}\right] r\mathrm{d}r\mathrm{d}z = 2\pi \oint_{\partial\Omega} (v_r n_r + v_z n_z)\mathrm{d}s, \tag{5.24}$$

where $2\pi ds = dS$. Now let us introduce a generic differentiable scalar field $a(r,z)$. A coordinate-free version of integration by parts is derived with use of (5.22) and the product rule for the gradient operator acting on $a\boldsymbol{v}$:

$$\int_\Omega a\nabla\cdot\boldsymbol{v}\,d\Omega = \oint_{\partial\Omega} a\boldsymbol{v}\cdot\boldsymbol{n}\,d\partial\Omega - \int_\Omega \nabla a\cdot\boldsymbol{v}\,d\Omega. \tag{5.25}$$

In component form for axisymmetric domains, the integration by parts identity in (5.25) reduces to

$$\begin{aligned}\int_\Omega a\left[\frac{1}{r}\frac{\partial(rv_r)}{\partial r}+\frac{\partial v_z}{\partial z}\right]r\,dr\,dz &= \oint_{\partial\Omega} a(v_r n_r + v_z n_z)\,ds \\ &\quad - \int_\Omega\left(\frac{\partial a}{\partial r}v_r + \frac{\partial a}{\partial z}v_z\right)r\,dr\,dz.\end{aligned} \tag{5.26}$$

Now we return to the strong form of governing partial differential equation (PDE) for steady Fourier heat conduction, (5.18), repeated here with heat source f moved to the left side:

$$\nabla\cdot(\boldsymbol{k}\nabla T) + f = 0, \qquad (r,\theta,z)\in\Omega. \tag{5.27}$$

A differentiable but otherwise arbitrary scalar field $w(r,z)$ is now prescribed over the domain. Multiplying (5.27) by w, integrating over Ω, and dividing by 2π yields

$$\int_\Omega [w\nabla\cdot(\boldsymbol{k}\nabla T)]r\,dr\,dz + \int_\Omega wf\,dr\,dz = 0. \tag{5.28}$$

Next, the identity in (5.25) is applied to the first term on the left side of (5.28) to produce

$$\int_\Omega w[\nabla\cdot(\boldsymbol{k}\nabla T)]r\,dr\,dz = \oint_{\partial\Omega}(w\boldsymbol{k}\nabla T)\cdot\boldsymbol{n}\,ds - \int_\Omega(\nabla w)\cdot(\boldsymbol{k}\nabla T)r\,dr\,dz. \tag{5.29}$$

Finally, substituting (5.29) back into (5.28), we arrive at the weak form PDE for steady-state Fourier conduction:

$$\int_\Omega(\nabla w)\cdot(\boldsymbol{k}\nabla T)r\,dr\,dz = \int_\Omega wf\,r\,dr\,dz + \oint_{\partial\Omega}(w\boldsymbol{k}\nabla T)\cdot\boldsymbol{n}\,ds. \tag{5.30}$$

Expressed in terms of cylindrical scalar components, (5.30) is expanded as follows after invoking (5.8) and (5.17):

$$\begin{aligned}\int_\Omega\left[\frac{\partial w}{\partial r}\left(k_{rr}\frac{\partial T}{\partial r}\right)+\frac{\partial w}{\partial z}\left(k_{zz}\frac{\partial T}{\partial z}\right)\right]r\,dr\,dz \\ = \int_\Omega wf\,r\,dr\,dz - \oint_{\partial\Omega} wq_n\,ds,\end{aligned} \tag{5.31}$$

recalling from §5.2 that the heat flux on the boundary (positive for outward energy flow) is

$$q_n = \boldsymbol{q}\cdot\boldsymbol{n} = q_r n_r + q_z n_z \qquad (s\in\partial\Omega). \tag{5.32}$$

The complete weak form for the problem statement for steady heat conduction in an axisymmetric domain seeks the temperature field $T(r,z)$ in domain Ω that satisfies first-order partial differential equation (5.31), given boundary conditions on $\partial\Omega$, imposed heat source f, and conductivity matrix $\boldsymbol{k}$ that may generally depend on coordinates r and z. The weak form governing equation contains first order partial derivatives of w and T. In contrast, the strong form PDE (5.19) contains second order derivatives of the temperature field $T(r,z)$. Notice the presence of the r multiplier acting on the term in square braces in the integrand on the left side of (5.31); no analogous term enters the weak form PDE for planar 2-D heat conduction derived in (4.17) of Chapter 4.

5.4 Heat Conduction: Matrix Equations

The weak form differential equation for steady heat conduction is now used to derive the finite element equilibrium equations for a single axisymmetric element. The weak form PDE is (5.31), which for a domain of integration Ω^e comprised of a single finite element with boundary $\partial\Omega^e$ becomes

$$\begin{aligned}\int_{\Omega^e}\left[\frac{\partial w}{\partial r}\left(k_{rr}\frac{\partial T}{\partial r}\right)+\frac{\partial w}{\partial z}\left(k_{zz}\frac{\partial T}{\partial z}\right)\right]r\mathrm{d}r\mathrm{d}z \\ =\int_{\Omega^e} wf\mathrm{d}r\mathrm{d}z-\oint_{\partial\Omega^e} wq_n\mathrm{d}s.\end{aligned} \tag{5.33}$$

Applying the summation convention for repeated indices, the temperature field $T(r,z)$ and the weight function field $w(r,z)$ are discretized as follows:

$$T(r,z)=N_j(r,z)T_j, \qquad w(r,z)=N_i(r,z)w_i, \tag{5.34}$$

where i or j runs from 1 to the total number of nodes or degrees of freedom for element e, and where T_j and w_i are discrete values of temperature T and w at node j or i with respective coordinates (r_j,z_j) or (r_i,z_i). Shape functions $N_i(r,z)$ for axisymmetric problems are identical to those introduced for planar 2-D problems in §4.3 upon making the substitutions $x\rightarrow r$ and $y\rightarrow z$.

Substituting (5.34) into (5.33) and factoring out the constants w_i results in

$$\begin{aligned}w_i\int_{\Omega^e}\left[\frac{\partial N_i}{\partial r}\left(k_{rr}\frac{\partial N_j}{\partial r}\right)+\frac{\partial N_i}{\partial z}\left(k_{zz}\frac{\partial N_j}{\partial z}\right)\right]T_j r\mathrm{d}r\mathrm{d}z \\ =w_i\int_{\Omega^e} N_i f r\mathrm{d}r\mathrm{d}z-w_i\oint_{\partial\Omega^e} N_i q_n\mathrm{d}s.\end{aligned} \tag{5.35}$$

Since the function w and hence its nodal values w_i are arbitrary, (5.35) leads to the following set of scalar element-level equilibrium equations, one for each nodal DOF or for each value of i:

$$\int_{\Omega^e}\left[\frac{\partial N_i}{\partial r}\left(k_{rr}\frac{\partial N_j}{\partial r}\right)+\frac{\partial N_i}{\partial z}\left(k_{zz}\frac{\partial N_j}{\partial z}\right)\right]T_j r\mathrm{d}r\mathrm{d}z$$
$$=\int_{\Omega^e}N_i f r\mathrm{d}r\mathrm{d}z-\oint_{\partial\Omega^e}N_i q_n \mathrm{d}s. \tag{5.36}$$

Factoring out the nodal constants T_j from the integral on the left side of the equality, all scalar equations in (5.36) can be written collectively in vector-matrix form as follows:

$$[\boldsymbol{k}^{(e)}]\{\boldsymbol{T}^{(e)}\}=\{\boldsymbol{f}^{(e)}\}+\{\boldsymbol{Q}^{(e)}\}, \tag{5.37}$$

where $\boldsymbol{T}^{(e)}$ is a column vector of nodal temperatures T_i. Nodal components of the heat source vector $\boldsymbol{f}^{(e)}$ and boundary term $\boldsymbol{Q}^{(e)}$ are defined as the following integrals, respectively:

$$f_i^{(e)}=\int_{\Omega^e}N_i(r,z)f(r,z)r\mathrm{d}r\mathrm{d}z, \qquad Q_i^{(e)}=-\oint_{\partial\Omega^e}N_i(s)q_n(s)\mathrm{d}s, \tag{5.38}$$

noting that on the element's boundary $\partial\Omega^e$, coordinates are parametrically described by functions $r=r(s)$ and $z=z(s)$ for any cross-sectional slice with $\theta=$ constant.

Components of the element stiffness for heat conduction (recalling the restriction to transverse or higher isotropy for Fourier conduction coefficients) are found via computation of the integral below, where $\theta=$ constant for any axisymmetric cross-section of the body:

$$k_{ij}^e=\int_{\Omega^e}\left[\frac{\partial N_i}{\partial r}\left(k_{rr}\frac{\partial N_j}{\partial r}\right)+\frac{\partial N_i}{\partial z}\left(k_{zz}\frac{\partial N_j}{\partial z}\right)\right]r\mathrm{d}r\mathrm{d}z. \tag{5.39}$$

Analogously to (4.33), components of the B-matrix for axisymmetric elements are defined as the following first derivatives, where here j runs from 1 to the total number of nodes n in the element:

$$B_{1j}=\frac{\partial N_j}{\partial r}, \qquad B_{2j}=\frac{\partial N_j}{\partial z}. \tag{5.40}$$

The element stiffness matrix of components in (5.39) can be expressed succinctly in vector-matrix notation as

$$[\boldsymbol{k}^{(e)}]=\int_{\Omega^e}[\boldsymbol{B}]^{\mathrm{T}}[\boldsymbol{k}][\boldsymbol{B}]r\mathrm{d}r\mathrm{d}z, \tag{5.41}$$

where $\boldsymbol{k}$ is the 2×2 diagonal thermal conductivity matrix of (5.16). The size of the element stiffness matrix is $n\times n$, where n is the total number of nodal degrees of freedom for the element, i.e., the number of nodes per element for conduction problems.

Derivations to this point in §5.4 apply for any axisymmetric element type (e.g., triangle, quadrilateral, etc.), where subscripts $i,j\geq 3$. Now consider linear triangular finite elements with shape functions listed in (4.25), where here $x\rightarrow r$ and $y\rightarrow z$. In vector-matrix form these shape functions are collected as

$$[\boldsymbol{N}(r,z)] = \frac{1}{2A_e}[\alpha_1 + \beta_1 r + \gamma_1 z \quad \alpha_2 + \beta_2 r + \gamma_2 z \quad \alpha_3 + \beta_3 r + \gamma_3 z]. \tag{5.42}$$

The area A_e is the enclosed area of the corresponding edges of the triangle represented at constant θ. The B-matrix (and its transpose) for linear triangular elements is the 2×3 matrix (and 3×2 matrix) of constants

$$[\boldsymbol{B}] = \frac{1}{2A_e}\begin{bmatrix} \beta_1 & \beta_2 & \beta_3 \\ \gamma_1 & \gamma_2 & \gamma_3 \end{bmatrix}, \qquad [\boldsymbol{B}]^{\mathrm{T}} = \frac{1}{2A_e}\begin{bmatrix} \beta_1 & \gamma_1 \\ \beta_2 & \gamma_2 \\ \beta_3 & \gamma_3 \end{bmatrix}. \tag{5.43}$$

For the particularly simple constitutive model of an isotropic and homogeneous Fourier heat conductor corresponding to the equilibrium conditions in (5.20), where k is the scalar conductivity, the stiffness matrix in (5.41) for linear triangular elements obeys

$$k_{ij}^{(e)} = \frac{k}{4A_e^2}(\beta_i\beta_j + \gamma_i\gamma_j)\int_{\Omega^e} r\mathrm{d}r\mathrm{d}z, \qquad (i,j = 1,2,3). \tag{5.44}$$

Unlike the planar 2-D case in (4.37), here the stiffness matrix is not an exclusive function of the coefficients (β_i, γ_i), area A_e, and conductivity k due to the presence of the r multiplier in the integrand of (5.44).

Natural boundary conditions for axisymmetric problems are treated in a similar manner as those for planar 2-D problems discussed in §4.5, albeit with the main caveat that the axisymmetric surface element $\mathrm{d}S = 2\pi\mathrm{d}s$ has an implicit dependence on $\mathrm{d}\theta$. For example, on the lateral surface of a right circular cylinder, $\mathrm{d}S = r\mathrm{d}\theta\mathrm{d}z$, while on the ends of such a cylinder, $\mathrm{d}S = r\mathrm{d}r\mathrm{d}\theta$. The θ-dependence is removed automatically by the following consideration: since fields are independent of θ for axisymmetric problems, area integration (with element $\mathrm{d}S$) over the domain $\theta \in (-\pi, \pi]$ reduces to line integration (with element $\mathrm{d}s$) over the slice of the boundary $\partial\Omega$ on any rz-plane at constant cross-section θ. Issues regarding assembly of the global equations from element equations, condensation of the global equations according to imposed boundary conditions, solution of the condensed system of global equations, and post-processing of numerical results for axisymmetric steady heat conduction all parallel the discussion of such topics for planar 2-D problems given in §4.6 of Chapter 4.

5.5 Axisymmetric Elasticity: Governing Equations

Axisymmetric elasticity problems are analyzed via finite elements in a similar manner as planar 2-D problems, where solution variables generally depend on radial and axial, r and z, coordinates in the former and x and y coordinates in the latter. A source of additional overhead in axisymmetric problems is that the strain-displacement relationships are more complex than in planar 2-D problems.

Governing equations from continuum mechanics and linear elasticity in cylindrical coordinates for axisymmetric domains are now presented. Only essential aspects are discussed; for more in-depth treatments, the reader is referred to books on continuum mechanics and elasticity, e.g., [1, 3]. As was the case in Chapter 4, the present scope is restricted to classical linear elasticity, wherein displacements and displacement gradients are presumed small in magnitude, and wherein stress is a linear function of strain.

In axisymmetric solid mechanics problems, the displacement vector $\boldsymbol{u}$ consists of two in-plane components at constant θ:

$$\{\boldsymbol{u}(r,z)\} = \begin{Bmatrix} u_1(x_1,x_2) \\ u_2(x_1,x_2) \end{Bmatrix} = \begin{Bmatrix} u_r(r,z) \\ u_z(r,z) \end{Bmatrix} = \begin{Bmatrix} u(r,z) \\ \upsilon(r,z) \end{Bmatrix}, \tag{5.45}$$

where the (u,υ) notation will be used most frequently in what follows. The symmetric strain tensor is defined as usual in coordinate-free tensor notation as

$$\boldsymbol{\varepsilon} = \frac{1}{2}[\nabla \boldsymbol{u} + (\nabla \boldsymbol{u})^{\mathrm{T}}]. \tag{5.46}$$

The standard description of linear elasticity [1, 3] defines strain in dimensionless physical components [2] as follows for axisymmetric problems:

$$\varepsilon_{rr} = \frac{\partial u}{\partial r}, \qquad \varepsilon_{zz} = \frac{\partial \upsilon}{\partial z}, \qquad \varepsilon_{\theta\theta} = \frac{u}{r}, \qquad \varepsilon_{rz} = \frac{1}{2}\left(\frac{\partial u}{\partial z} + \frac{\partial \upsilon}{\partial r}\right). \tag{5.47}$$

The other two strain components that may emerge with nonzero values in 3-D cylindrical coordinates, specifically $\varepsilon_{r\theta}$ and $\varepsilon_{\theta z}$, vanish identically by construction for axisymmetric problems. The four strain tensor components of (5.47) are often collected in vector form as

$$\begin{Bmatrix} \varepsilon_{rr} \\ \varepsilon_{zz} \\ \varepsilon_{\theta\theta} \\ \gamma_{rz} \end{Bmatrix} = \begin{Bmatrix} \varepsilon_{rr} \\ \varepsilon_{zz} \\ \varepsilon_{\theta\theta} \\ 2\varepsilon_{rz} \end{Bmatrix} = \begin{Bmatrix} \partial u/\partial r \\ \partial \upsilon/\partial z \\ u/r \\ \partial u/\partial z + \partial \upsilon/\partial r \end{Bmatrix}. \tag{5.48}$$

Compatibility conditions that ensure that a displacement vector exists producing a given strain field impose restrictions on derivatives of strain. For axisymmetric problems, compatibility of a given strain field demands [1, 3]

$$\frac{\partial^2 \varepsilon_{rr}}{\partial z^2} + \frac{\partial^2 \varepsilon_{zz}}{\partial r^2} = \frac{\partial^2 \gamma_{rz}}{\partial r \partial z}. \tag{5.49}$$

The conservation law to be solved in a static continuum mechanics problem is the local balance of linear momentum, expressed first here in coordinate-free tensor notation:

$$\nabla \cdot \boldsymbol{\sigma} + \boldsymbol{f} = 0. \tag{5.50}$$

The stress tensor is $\boldsymbol{\sigma}$; the body force vector (force per unit volume) is $\boldsymbol{f}$. As noted in Chapter 4, perhaps the most common example of a body force is that due to gravity. Note that for (quasi-)static problems, no inertial terms involving the product of mass density and local acceleration are included. The local balance of angular momentum results in the requirement that the stress tensor be symmetric, i.e., $\sigma_{ij} = \sigma_{ji}$.

For axisymmetric problems with physical components of stress, (5.50) consists of two independent equations that are obtained from (5.12) and the symmetry of tensor $\boldsymbol{\sigma}$:

$$\frac{\partial \sigma_{rr}(r,z)}{\partial r} + \frac{1}{r}[\sigma_{rr}(r,z) - \sigma_{\theta\theta}(r,z)] + \frac{\partial \sigma_{rz}(r,z)}{\partial z} + f_r(r,z) = 0, \tag{5.51}$$

$$\frac{\partial \sigma_{rz}(r,z)}{\partial r} + \frac{\sigma_{rz}(r,z)}{r} + \frac{\partial \sigma_{zz}(r,z)}{\partial z} + f_z(r,z) = 0. \tag{5.52}$$

The constitutive equations of linear elasticity, which may be isotropic or more generally transversely isotropic for axisymmetric problems, apply, and the following discussion identically parallels that of §4.7. Stress is linearly related to strain via the matrix/tensor of elastic constants $\boldsymbol{C}$. In full index notation for a general 3-D problem,

$$\sigma_{ij} = C_{ijkl}\varepsilon_{kl}, \qquad (i,j,k,l = 1,2,3). \tag{5.53}$$

An equivalent constitutive equation is often written using Voigt notation [3]:

$$\{\boldsymbol{\sigma}\} = [\boldsymbol{C}]\{\boldsymbol{\varepsilon}\} \leftrightarrow \sigma_I = C_{IJ}\varepsilon_J, \qquad (I,J = 1,2,\ldots 6). \tag{5.54}$$

In (5.54), $\boldsymbol{C}$ is a symmetric 6×6 matrix relating column vectors of stress and strain. For the most general class of anisotropic material, $\boldsymbol{C}$ consists of a maximum of 21 independent constants; material symmetries reduce this number to a minimum of 2 elastic constants for isotropy. The strain energy density W is the following quadratic form with corresponding first and second derivatives:

$$W = \frac{1}{2}C_{ijkl}\varepsilon_{ij}\varepsilon_{kl} \Rightarrow \sigma_{ij} = \frac{\partial W}{\partial \varepsilon_{ij}}, \qquad C_{ijkl} = \frac{\partial \sigma_{ij}}{\partial \varepsilon_{kl}} = \frac{\partial^2 W}{\partial \varepsilon_{ij}\partial \varepsilon_{kl}}. \tag{5.55}$$

It follows that the elasticity tensor has the symmetries $C_{ijkl} = C_{jikl}$, $C_{ijkl} = C_{ijlk}$, and $C_{klij} = C_{ijkl}$ that enable a reduction from $3^4 = 81$ to the 21 independent components mentioned above.

We consider an axisymmetric domain Ω with boundary $\partial\Omega$, where cylindrical coordinates (r,θ,z) cover the domain, and where field variables do not depend on angular coordinate θ. The unit outward normal vector to Ω again is denoted by $\boldsymbol{n}(r,z)$. Boundary conditions on $\partial\Omega$ are of two possible types:

- **Essential**: displacements (u, υ) prescribed on $\partial\Omega_u$;
- **Natural**: traction (t_r,t_z) prescribed on $\partial\Omega_t$.

The mechanical traction vector on the boundary is defined in physical components as

$$\boldsymbol{t} = \boldsymbol{\sigma}\cdot\boldsymbol{n} = \boldsymbol{n}\cdot\boldsymbol{\sigma} \leftrightarrow t_i = \sigma_{ij}n_j = n_j\sigma_{ji}. \tag{5.56}$$

It is also possible to invoke mixed boundary conditions, e.g., imposed (u, t_z) or (t_r, υ), but imposition of both entries of a work conjugate pair (u, t_r) or (υ, t_z) at the same location on $\partial\Omega$ is physically unacceptable.

More specific constitutive laws apply for isotropic materials as discussed next for axisymmetric linear elastic solid bodies.

5.5.1 Isotropic Linear Elasticity

For isotropic elastic solids in the present axisymmetric coordinate system, stress components obey

$$\sigma_{r\theta} = \sigma_{\theta z} = 0, \qquad (\text{generally, } \sigma_{\theta\theta} \neq 0). \tag{5.57}$$

Recall also that circumferential or hoop strain $\varepsilon_{\theta\theta} = u/r$ is generally nonzero. In isotropic linear elasticity, the four generally nonzero physical components of stress and strain are related by the constitutive law

$$\begin{Bmatrix} \sigma_{rr} \\ \sigma_{zz} \\ \sigma_{\theta\theta} \\ \sigma_{rz} \end{Bmatrix} = \begin{bmatrix} C_{11} & C_{12} & C_{12} & 0 \\ C_{12} & C_{11} & C_{12} & 0 \\ C_{12} & C_{12} & C_{11} & 0 \\ 0 & 0 & 0 & C_{66} \end{bmatrix} \begin{Bmatrix} \varepsilon_{rr} \\ \varepsilon_{zz} \\ \varepsilon_{\theta\theta} \\ \gamma_{rz} \end{Bmatrix} \leftrightarrow \{\boldsymbol{\sigma}\} = [\boldsymbol{C}]\{\boldsymbol{\varepsilon}\}. \tag{5.58}$$

Introducing the usual Young's modulus E and Poisson's ratio ν, the first of (5.58) reduces to

$$\begin{Bmatrix} \sigma_{rr} \\ \sigma_{zz} \\ \sigma_{\theta\theta} \\ \sigma_{rz} \end{Bmatrix} = \frac{E(1-\nu)}{(1+\nu)(1-2\nu)} \begin{bmatrix} 1 & \frac{\nu}{1-\nu} & \frac{\nu}{1-\nu} & 0 \\ \frac{\nu}{1-\nu} & 1 & \frac{\nu}{1-\nu} & 0 \\ \frac{\nu}{1-\nu} & \frac{\nu}{1-\nu} & 1 & 0 \\ 0 & 0 & 0 & \frac{1-2\nu}{2-2\nu} \end{bmatrix} \begin{Bmatrix} \varepsilon_{rr} \\ \varepsilon_{zz} \\ \varepsilon_{\theta\theta} \\ \gamma_{rz} \end{Bmatrix}. \tag{5.59}$$

5.5.2 Strong Form

The strong form PDEs are derived as follows for isotropic linear elasticity in axisymmetric domains under quasi-static conditions. Here we also assume that elastic coefficients are all constants, independent of location (r, θ, z) in the solid, meaning that the material is elastically homogeneous. The linear elastic constitutive equations in (5.58) are first substituted into equilibrium equations (5.51) and (5.52) to produce, respectively,

$$\frac{\partial}{\partial r}[C_{11}\varepsilon_{rr}+C_{12}(\varepsilon_{zz}+\varepsilon_{\theta\theta})]+\frac{\partial}{\partial z}[C_{66}\gamma_{rz}] + \frac{1}{r}[(C_{11}-C_{12})(\varepsilon_{rr}-\varepsilon_{\theta\theta})]+f_r=0, \tag{5.60}$$

$$\frac{\partial}{\partial z}[C_{11}\varepsilon_{zz}+C_{12}(\varepsilon_{rr}+\varepsilon_{\theta\theta})]+\frac{\partial}{\partial r}[C_{66}\gamma_{rz}]+\frac{1}{r}[C_{66}\gamma_{rz}]+f_z=0. \tag{5.61}$$

Next, we introduce the following equivalent and alternative set of two isotropic elastic constants, specifically the Lamé modulus λ and the shear modulus G:

$$\lambda=\frac{E\nu}{(1+\nu)(1-2\nu)}=C_{12}, \qquad G=\frac{E}{2(1+\nu)}=C_{66}. \tag{5.62}$$

We also define the continuum dilatation field $e(r,z)$ and the continuum rotation field $\omega(r,z)$ as, respectively,

$$e=\varepsilon_{rr}+\varepsilon_{\theta\theta}+\varepsilon_{zz}=\frac{1}{r}\frac{\partial}{\partial r}(ru)+\frac{\partial \upsilon}{\partial z}, \qquad \omega=\frac{1}{2}\left(\frac{\partial u}{\partial z}-\frac{\partial \upsilon}{\partial r}\right). \tag{5.63}$$

Finally, after substituting from the stress-strain relations in (5.59) and using definitions in (5.62) and (5.63), momentum conservation equations (5.60) and (5.61) can be expressed as [1]

$$(\lambda+2G)\frac{\partial e}{\partial r}+2G\frac{\partial \omega}{\partial z}+f_r=0, \tag{5.64}$$

$$(\lambda+2G)\frac{\partial e}{\partial z}-\frac{2G}{r}\frac{\partial}{\partial r}(r\omega)+f_z=0. \tag{5.65}$$

The complete strong form for the problem statement seeks the displacement field $[u(r,z),\upsilon(r,z)]$ in domain Ω that satisfies the second-order partial differential equations (5.64) and (5.65), given boundary conditions on $\partial\Omega$, imposed body force field f, and elastic constants E and ν (or equivalently, λ and G) for a homogeneous isotropic solid.

5.6 Elasticity: Weak Form

The weak form of the problem statement for axisymmetric linear elasticity problems is derived from the strong form following a procedure analogous to that for 2-D elasticity problems considered already in §4.8. Recall that the present derivation is restricted to a quasi-static and linear elastic response, where the entire problem domain is Ω and its boundary is $\partial\Omega$ with unit outward normal vector $\boldsymbol{n}$. As noted in §5.5.1, maximal material symmetry of the elasticity tensor used in subsequent derivations also requires that the solid is isotropic with only two independent elastic constants.

The derivation of the weak form PDEs again begins with introduction of a differentiable but otherwise arbitrary vector-valued weight function $\boldsymbol{w}(r,z)$ with two components for axisymmetric analyses:

$$\{\boldsymbol{w}(\boldsymbol{x})\} = \begin{Bmatrix} w_1(x_1,x_2) \\ w_2(x_1,x_2) \end{Bmatrix} = \begin{Bmatrix} w_r(r,z) \\ w_z(r,z) \end{Bmatrix}. \tag{5.66}$$

The derivation proceeds most efficiently when (coordinate-free) tensor notation is invoked. Taking the scalar product of $\boldsymbol{w}$ with the local linear momentum balance of (5.50) and integrating over the problem domain produces

$$\int_{\Omega} \boldsymbol{w}\cdot(\nabla\cdot\boldsymbol{\sigma})\mathrm{d}\Omega + \int_{\Omega} \boldsymbol{w}\cdot\boldsymbol{f}\mathrm{d}\Omega = 0. \tag{5.67}$$

Integrating the first term on the left by parts with application of the divergence theorem, i.e., the integral theorem of Gauss discussed in §4.2 and §5.3, gives the identity

$$\begin{aligned} \int_{\Omega} \boldsymbol{w}\cdot(\nabla\cdot\boldsymbol{\sigma})\mathrm{d}\Omega &= \int_{\Omega} \nabla\cdot(\boldsymbol{w}\cdot\boldsymbol{\sigma})\mathrm{d}\Omega - \int_{\Omega}(\nabla\boldsymbol{w}):\boldsymbol{\sigma}\mathrm{d}\Omega \\ &= \oint_{\partial\Omega} \boldsymbol{w}\cdot\boldsymbol{\sigma}\cdot\boldsymbol{n}\,\mathrm{d}\partial\Omega - \int_{\Omega}(\nabla\boldsymbol{w}):\boldsymbol{\sigma}\mathrm{d}\Omega. \end{aligned} \tag{5.68}$$

The scalar product of two second-order tensors is, in index notation,

$$(\nabla\boldsymbol{w}):\boldsymbol{\sigma} = \nabla_j w_i \sigma_{ij}, \tag{5.69}$$

where physical components are implied but are not necessary. Substituting (5.68) into (5.67) and using the definition of the mechanical traction vector from (5.56), we arrive at the following integral PDE:

$$\int_{\Omega}(\nabla\boldsymbol{w}):\boldsymbol{\sigma}\mathrm{d}\Omega - \oint_{\partial\Omega} \boldsymbol{w}\cdot\boldsymbol{t}\mathrm{d}\partial\Omega - \int_{\Omega} \boldsymbol{w}\cdot\boldsymbol{f}\mathrm{d}\Omega = 0. \tag{5.70}$$

Now invoking linear elasticity constitutive equation $\boldsymbol{\sigma} = \boldsymbol{C}:\boldsymbol{\varepsilon}$ from (5.58), (5.70) can be expressed as

$$\int_{\Omega}(\nabla\boldsymbol{w}):\boldsymbol{C}:\boldsymbol{\varepsilon}\mathrm{d}\Omega - \oint_{\partial\Omega} \boldsymbol{w}\cdot\boldsymbol{t}\mathrm{d}\partial\Omega - \int_{\Omega} \boldsymbol{w}\cdot\boldsymbol{f}\mathrm{d}\Omega = 0. \tag{5.71}$$

In indicial notation with physical components in axisymmetric cylindrical domains, this can be written as follows upon taking advantage of the symmetry of $\boldsymbol{C}$, where capital indices $I,J = 1,2,3,4 \leftrightarrow ()_{rr}, ()_{zz}, ()_{\theta\theta}, ()_{rz}$ and lower-case indices $i = 1,2 \leftrightarrow ()_r, ()_z$:

$$\int_{\Omega}(\nabla w_I)C_{IJ}\varepsilon_J r\mathrm{d}r\mathrm{d}z - \oint_{\partial\Omega} w_i t_i \mathrm{d}s - \int_{\Omega} w_i f_i r\mathrm{d}r\mathrm{d}z = 0. \tag{5.72}$$

The symmetric gradient vector of weight functions entering (5.72) is defined in physical components analogously to the strain vector in axisymmetric problems of (5.48):

$$\begin{Bmatrix} (\nabla w)_{rr} \\ (\nabla w)_{zz} \\ (\nabla w)_{\theta\theta} \\ (\nabla w)_{rz} + (\nabla w)_{zr} \end{Bmatrix} = \begin{Bmatrix} \partial w_r / \partial r \\ \partial w_z / \partial z \\ w_r / r \\ \partial w_r / \partial z + \partial w_z / \partial r \end{Bmatrix}. \tag{5.73}$$

Equations (5.71) and (5.72) are scalar representations, in coordinate-free notation and physical cylindrical coordinates, respectively, of the weak form PDE for quasi-static linear elasticity. For axisymmetric problems, a weak form problem statement seeks the displacement field $\boldsymbol{u}(\boldsymbol{x}) = [u(r,z), \upsilon(r,z)]$ in domain Ω that satisfies these partial differential equations (5.71) or (5.72), given boundary conditions on $\partial\Omega$, imposed body force field $\boldsymbol{f}$, and the elasticity matrix $\boldsymbol{C}$, restricted here to isotropic material symmetry. Either of these governing PDEs contains only first derivatives of $\boldsymbol{u}$ and $\boldsymbol{w}$, in contrast to the strong form PDEs in (5.60) and (5.61) that contain second derivatives of $\boldsymbol{u}$ with respect to r and z. In §5.7, upon introduction of shape functions and subsequent factorization, multiple or a vector-valued set of equilibrium equations will be extracted from the single scalar equation for the weak form derived above.

5.7 Elasticity: Matrix Equations

Derivation of the finite element equilibrium equations for axisymmetric elasticity is very similar to derivation of the analogous equilibrium equations for axisymmetric heat conduction. Hence, the steps described next parallel those of §5.4. The same general classes of 2-D/axisymmetric interpolation functions $N_i(r,z)$ introduced in §4.3 are used here, as are particular shape functions for linear triangular elements in some examples that follow. Recall from §5.4 that the substitutions $x \to r$ and $y \to z$ are used to transform planar 2-D shape functions of §4.3 to axisymmetric interpolation/shape functions.

The weak form differential equations for quasi-static linear elasticity, restricted to isotropic materials, are now used to derive the finite element equilibrium equation for a single element. The primary solution variable is a two-entry vector field [displacement $\boldsymbol{u}(r,z)$], so the number of degrees of freedom per node is thus two: $(u_r, u_z) = (u, \upsilon)$.

A weak form PDE for axisymmetric static linear elasticity is (5.72), which for a domain of integration Ω^e consisting of a single element becomes

$$\int_{\Omega^e} (\nabla w_I) C_{IJ} \varepsilon_J r \mathrm{d}r \mathrm{d}z = \oint_{\partial\Omega^e} w_i t_i \mathrm{d}s + \int_{\Omega^e} w_i f_i r \mathrm{d}r \mathrm{d}z = 0. \tag{5.74}$$

Applying the summation convention over repeated indices, displacement fields $u(r,z)$ and $\upsilon(r,z)$ and weight function fields $w_r(r,z)$ and $w_z(r,z)$ are discretized as follows:

$$\begin{Bmatrix} u(r,z) \\ \upsilon(r,z) \end{Bmatrix} = \begin{Bmatrix} N_j(r,z) u_j \\ N_j(r,z) \upsilon_j \end{Bmatrix}, \qquad \begin{Bmatrix} w_r(r,z) \\ w_z(r,z) \end{Bmatrix} = \begin{Bmatrix} N_i(r,z) w_{ri} \\ N_i(r,z) w_{zi} \end{Bmatrix}. \tag{5.75}$$

Subscripts i or j span values of 1 to the total number of nodes n in element e, and u_j and υ_j are nodal values of displacements at node j with cylindrical coordinates (r_j, z_j) at any fixed value of θ. Also, (w_{ri}, w_{zi}) are discrete values of (w_r, w_z) at node i with coordinates (r_i, z_i). These discrete values will later be alternatively written as w_{ki}, where k corresponds to r or z for axisymmetric problems. The discretized displacement functions can be written more concisely as

$$\{\boldsymbol{u}\} = \begin{Bmatrix} u \\ \upsilon \end{Bmatrix} = [\boldsymbol{N}]\{\boldsymbol{u}^{(e)}\} = \begin{bmatrix} N_1 & 0 & N_2 & 0 & N_3 & 0 & \dots \\ 0 & N_1 & 0 & N_2 & 0 & N_3 & \dots \end{bmatrix} \begin{Bmatrix} u_1 \\ \upsilon_1 \\ u_2 \\ \upsilon_2 \\ u_3 \\ \upsilon_3 \\ \vdots \end{Bmatrix}, \tag{5.76}$$

with $\boldsymbol{N}$ a $2 \times 2n$ matrix of shape functions padded with zeros as indicated after the final equality. The B-matrix for axisymmetric linear elasticity is defined as the following $4 \times 2n$ matrix of interpolation functions and their first partial coordinate derivatives:

$$[\boldsymbol{B}] = \begin{bmatrix} \frac{\partial N_1}{\partial r} & 0 & \frac{\partial N_2}{\partial r} & 0 & \frac{\partial N_3}{\partial r} & 0 & \dots \\ 0 & \frac{\partial N_1}{\partial z} & 0 & \frac{\partial N_2}{\partial z} & 0 & \frac{\partial N_3}{\partial z} & \dots \\ \frac{N_1}{r} & 0 & \frac{N_2}{r} & 0 & \frac{N_3}{r} & 0 & \dots \\ \frac{\partial N_1}{\partial z} & \frac{\partial N_1}{\partial r} & \frac{\partial N_2}{\partial z} & \frac{\partial N_2}{\partial r} & \frac{\partial N_3}{\partial z} & \frac{\partial N_3}{\partial r} & \dots \end{bmatrix}. \tag{5.77}$$

This definition enables computation of the strain vector in (5.48) at any location within the element e from its vector of nodal displacements $\boldsymbol{u}^{(e)}$ as

$$\{\boldsymbol{\varepsilon}^{(e)}(r,z)\} = \begin{Bmatrix} \varepsilon_{rr}(r,z) \\ \varepsilon_{zz}(r,z) \\ \varepsilon_{\theta\theta}(r,z) \\ \gamma_{rz}(r,z) \end{Bmatrix} = [\boldsymbol{B}(r,z)]\{\boldsymbol{u}^{(e)}\}. \tag{5.78}$$

A completely analogous equation holds for interpolation of the symmetrized gradient vector of weight functions defined in (5.73).

We now return to the weak form integro-differential equation. Substituting (5.75) into (5.74) and factoring out the constants w_{im} [recalling $w_i(r,z) = N_m(r,z)w_{im}$ with i referring to r or z and m referring to a local node number] and using (5.78) results in the following discretized expression in vector-matrix form:

$$[\boldsymbol{w}^{(e)}]^{\mathrm{T}} \cdot \left[\int_{\Omega^e} [\boldsymbol{B}]^{\mathrm{T}} [\boldsymbol{C}][\boldsymbol{B}]\{\boldsymbol{u}^{(e)}\} r \mathrm{d}r \mathrm{d}z \right] = [\boldsymbol{w}^{(e)}]^{\mathrm{T}} \cdot \int_{\Omega^e} [\boldsymbol{N}]^{\mathrm{T}} \{\boldsymbol{f}\} r \mathrm{d}r \mathrm{d}z + [\boldsymbol{w}^{(e)}]^{\mathrm{T}} \cdot \oint_{\partial \Omega^e} [\boldsymbol{N}]^{\mathrm{T}} \{\boldsymbol{t}\} \mathrm{d}s. \tag{5.79}$$

Nodal values of weight functions w_r and w_z, i.e., values w_{ij}, are collected in the column vector $\boldsymbol{w}^{(e)}$, which, like $\boldsymbol{u}^{(e)}$, is of size $2n$. Since functions $[w_r(r,z), w_z(r,z)]$ and hence w_{ij} and $\boldsymbol{w}^e$ are arbitrary, it follows that (5.79) leads to the following set of $m \times i = 2n$ scalar equations, two for each value of m or two for each node:

$$\left[\int_{\Omega^e} [\boldsymbol{B}]^{\mathrm{T}} [\boldsymbol{C}][\boldsymbol{B}] r \mathrm{d}r \mathrm{d}z \right] \{\boldsymbol{u}^{(e)}\} = \int_{\Omega^e} [\boldsymbol{N}]^{\mathrm{T}} \{\boldsymbol{f}\} r \mathrm{d}r \mathrm{d}z + \oint_{\partial \Omega^e} [\boldsymbol{N}]^{\mathrm{T}} \{\boldsymbol{t}\} \mathrm{d}s. \tag{5.80}$$

Nodal displacements $\boldsymbol{u}^{(e)}$ have been factored from the integral term on the left side since they are constant over element domain Ω_e. In symbolic form, (5.80) is written concisely as the familiar

$$[\boldsymbol{k}^{(e)}]\{\boldsymbol{u}^{(e)}\} = \{\boldsymbol{f}^{(e)}\} + \{\boldsymbol{Q}^{(e)}\} = \{\boldsymbol{F}^{(e)}\}, \tag{5.81}$$

where $\boldsymbol{k}^{(e)}$ is the $2n \times 2n$ element stiffness matrix defined according to

$$[\boldsymbol{k}^{(e)}] = \int_{\Omega^e} [\boldsymbol{B}]^{\mathrm{T}} [\boldsymbol{C}][\boldsymbol{B}] r \mathrm{d}r \mathrm{d}z. \tag{5.82}$$

Also entering (5.81) are $\boldsymbol{f}^{(e)}$, the body force vector resolved at the nodes, and $\boldsymbol{Q}^{(e)}$, the contribution of natural boundary conditions (i.e., applied tractions) resolved at the nodes. The latter two column vectors, and the their sum, the total load vector $\boldsymbol{F}^{(e)}$, are each of length $2n$. Referring to terms in (5.80) applied here for axisymmetric and isotropic linear elasticity, $\boldsymbol{C}$ is a symmetric 4×4 matrix of the equivalent form in (5.58) or (5.59), $\boldsymbol{B}$ is a $4 \times 2n$ matrix of the form in (5.77), $\boldsymbol{N}$ is a $2 \times 2n$ matrix of the form in (5.76), $\boldsymbol{f}$ is simply the body force vector with physical components (f_r, f_z), and $\boldsymbol{t}$ is simply the traction vector of (5.56) with physical components (t_r, t_z).

Derivations for finite element equilibrium equations of linear elasticity up to this point in §5.7 apply for any axisymmetric element referred to cylindrical coordinates. Now consider linear triangular finite elements with shape functions listed in (4.25) and with $(x, y) \to (r, z)$. In vector-matrix form, polynomial interpolation functions of (5.76) and their derivatives in (5.77) are, for this element type,

$$[\boldsymbol{N}(r,z)]^{\mathrm{T}} = \frac{1}{2A_e} \begin{bmatrix} \alpha_1 + \beta_1 r + \gamma_1 z & 0 \\ 0 & \alpha_1 + \beta_1 r + \gamma_1 z \\ \alpha_2 + \beta_2 r + \gamma_2 z & 0 \\ 0 & \alpha_2 + \beta_2 r + \gamma_2 z \\ \alpha_3 + \beta_3 r + \gamma_3 z & 0 \\ 0 & \alpha_3 + \beta_3 r + \gamma_3 z \end{bmatrix}; \tag{5.83}$$

$$[\boldsymbol{B}(r,z)] = \frac{1}{2A_e}\begin{bmatrix} \beta_1 & 0 & \beta_2 & 0 & \beta_3 & 0 \\ 0 & \gamma_1 & 0 & \gamma_2 & 0 & \gamma_3 \\ \frac{\alpha_1}{r}+\beta_1+\frac{\gamma_1 z}{r} & 0 & \frac{\alpha_2}{r}+\beta_2+\frac{\gamma_2 z}{r} & 0 & \frac{\alpha_3}{r}+\beta_3+\frac{\gamma_3 z}{r} & 0 \\ \gamma_1 & \beta_1 & \gamma_2 & \beta_2 & \gamma_3 & \beta_3 \end{bmatrix}, \quad (5.84)$$

$$[\boldsymbol{B}(r,z)]^{\mathrm{T}} = \frac{1}{2A_e}\begin{bmatrix} \beta_1 & 0 & \frac{\alpha_1}{r}+\beta_1+\frac{\gamma_1 z}{r} & \gamma_1 \\ 0 & \gamma_1 & 0 & \beta_1 \\ \beta_2 & 0 & \frac{\alpha_2}{r}+\beta_2+\frac{\gamma_2 z}{r} & \gamma_2 \\ 0 & \gamma_2 & 0 & \beta_2 \\ \beta_3 & 0 & \frac{\alpha_3}{r}+\beta_3+\frac{\gamma_3 z}{r} & \gamma_3 \\ 0 & \gamma_3 & 0 & \beta_3 \end{bmatrix}. \quad (5.85)$$

Here A_e is the area enclosed by the edges of the triangular element in the rz-plane at constant θ. Recall from §4.9 that in planar 2-D problems, $\boldsymbol{B}$ depends only on the area of the element A_e and the scalar coefficients γ_i and β_i, all of which are constants depending only on the nodal coordinates (x_i, y_i) from (4.24). In contrast, for axisymmetric problems, $\boldsymbol{B}$ contains terms of order $1/r$ and z/r. These terms manifest from the axisymmetric strain tensor's inclusion of the component $\varepsilon_{\theta\theta}$, often called the hoop strain, that is a function of $1/r$ that varies with the radial position r in the element. Due to the commensurate additional complexity associated with the presence of these non-constant terms and the factor of r in differential element $r\mathrm{d}r\mathrm{d}z$ entering the integrand of (5.82), closed-form expressions for stiffness matrix components $k_{ij}^{(e)}$ analogous to those derived in (4.91) and (4.92) cannot be obtained analytically for axisymmetric finite elements in solid mechanics. Rather, such element-level integrals are usually calculated in FEM software via numerical integration/quadrature routines.

A possibly important distinction between the stiffness matrix in planar 2-D elements and axisymmetric elements for linear elasticity is now mentioned. The B-matrix of (5.77) contains terms of order $1/r$ as already noted [see, e.g., (5.84) and (5.85) for the particular case of linear triangular elements], as does the integrand $\boldsymbol{B}^{\mathrm{T}}\boldsymbol{C}\boldsymbol{B}r\mathrm{d}r\mathrm{d}z$ in the element stiffness matrix, as evident from equation (5.82). Depending on limits of integration corresponding to the radial position of the element, the integrals that arise from (5.82) are classified as regular, singular, or nearly singular, where the latter two cases can be problematic and emerge for elements on or very close to the axis of symmetry at $r = 0$. For example, many standard numerical integration strategies are inaccurate for evaluating singular and nearly singular integrals that arise when an element lies on or close to the axis of rotation at $r = 0$. This issue is discussed further in the reference [5].

The remainder of the analysis of an axisymmetric, static linear elasticity problem using FEA follows the same general procedure as that for planar 2-D problems discussed already in Chapter 4. In particular, the key remaining steps in an analysis are assembly of the global equations from the element equations and mesh connectivity, imposition of boundary conditions and condensation of the global equations, solution of the global equilibrium equations, and post-processing. Most, if not all, of these often tedious tasks are performed efficiently by any modern FEM software package.

5.8 ANSYS Example: Hollow Cylindrical Column

The numerical example of this chapter demonstrates concepts from axisymmetric elasticity theory with the ANSYS software (v17.2, 2016). This problem is intentionally simple to enable validation of its numerical solution with an engineering approximation.

5.8.1 *Problem Statement*

Consider the cylindrical structure shown on the left of Fig. 5.2. The dimensions of the cylinder, which is hollow and capped and of uniform wall thickness $t = 2$ mm, are shown in the figure. The cylinder, which is essentially a metal can, is made of an aluminum alloy with Young's modulus $E = 69$ GPa, Poisson's ratio $\nu = 0.33$, and yield strength $\sigma_y = 276$ MPa. The dimensions of the cylinder are not drawn to scale. A uniform compressive pressure $p = 2$ MPa is enacted on the upper capped surface of the cylinder which rests on a frictionless surface. To reduce stress concentrations associated with sharp corners, the internal edges where caps intersect the lateral wall are filleted with radii of $t/2$.

The main objectives of this example are reduction of the 3-D system on the left side of Fig. 5.2 to an axisymmetric problem, determination of the maximum deflection of any point in the can, and calculation of the maximum effective stress at any point in the can. This stress is then to be compared with the yield stress of the material to consider whether plastic deformation would be expected. A final objective is validation of the results obtained from the FE software.

The first objective is achieved as follows, prior to use of the software. An axisymmetric rendering of the problem is shown on the right side of Fig. 5.2. The left sides of the half-cylinder, corresponding to $r = 0$, are constrained by symmetry boundary conditions to have null radial displacement, i.e., $u_r(r = 0) = 0$. These essential boundary conditions (BCs) prevent a hole from opening in the center and also prevent overlap of material in the radial direction. The lower surface of the cylinder is constrained to have null axial displacement: $u_z(z = 0) = 0$ where here z is the axial coordinate. The corner point that becomes fixed occupies the origin of the cylindrical coordinate system and ensures that no rigid body motion ensues. The pressure p, now acting along $z = h$, will be input as a natural BC later in ANSYS. The lateral surface at $r = R_0$ is traction free and permitted to expand, for example.

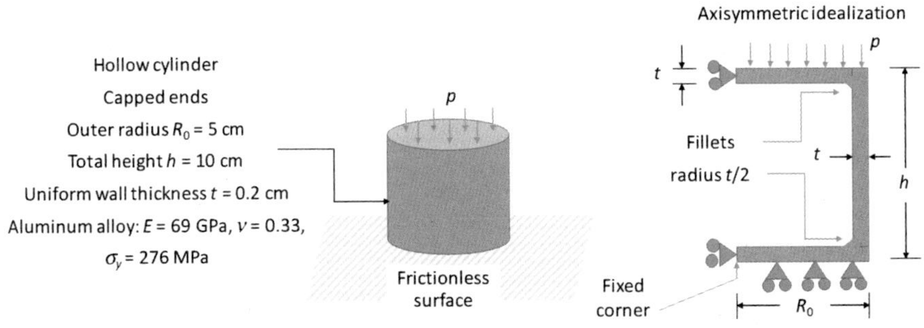

Fig. 5.2 Axisymmetric elasticity example problem: geometry, boundary conditions, and properties

5.8.2 Pre-processing

Pre-processing ensues by the following steps. The element type with prescription of axisymmetric conditions and material properties are input into ANSYS. This is followed by creation of the geometry, meshing, and enforcing of essential and natural boundary conditions. The element type for planar continuum elasticity, the `quad 4 node 182` element, is used:

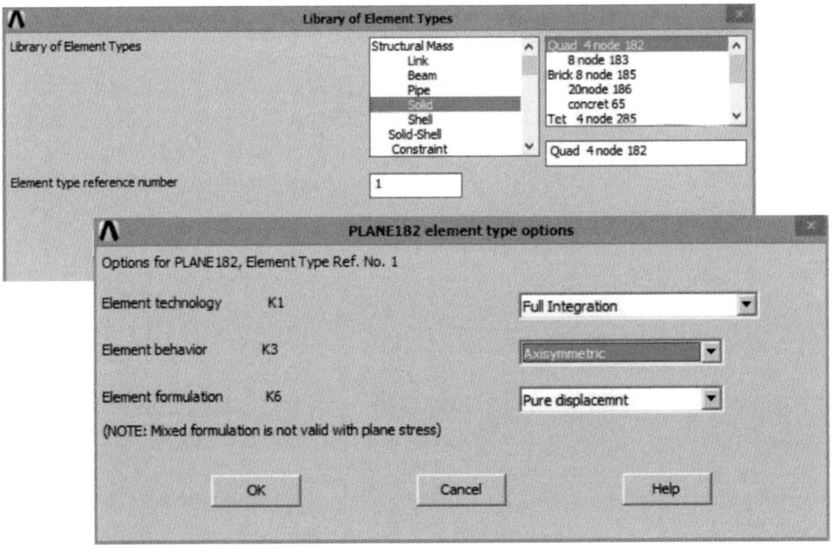

Fig. 5.3 Axisymmetric elasticity example problem: element data

```
Main Menu > Preprocessor > Element Type > Add/Edit/Delete
> Add > Quad 4 node 182
```

The choice is depicted in Fig. 5.3, wherein the element option for axisymmetry is also shown. The material model for the aluminum is isotropic linear elasticity with properties input as

```
Main Menu > Preprocessor > Material Properties > Material
Models > Structural > Linear > Elastic > Isotropic
```

The elastic modulus E and Poisson's ratio ν are respectively entered as 69e9 Pa and 0.33 in accordance with Fig. 5.2. The default coordinate system in ANSYS for axisymmetric problems uses the x-direction for the radial coordinate and the y-direction for the axial coordinate. In other words, the coordinate system used earlier in Chapter 5 of this book is recoverd via $x \to r$ and $y \to z$. A planar body in the xy-plane is created by generating three rectangles then adding them together. The fillets at the internal corners in Fig. 5.2 will be added subsequently. In the context of axisymmetry, this planar body is rotated 360 degrees about the y-axis to produce the cylindrical structure. Rectangles are produced by three iterations of the sequence

```
Main Menu > Preprocessor > Modeling > Create > Areas >
Rectangle > By Dimensions
```

Addition of the three areas to produce a single body is executed by

```
Main Menu > Preprocessor > Modeling > Operate > Booleans >
Add > Areas
```

Pick all three rectangles and then click "OK". Fillets are created at the two internal corners as indicated in Fig. 5.2. The procedure for creation of an area fillet is the same as that followed in §4.10. The command to initiate the process is

```
Main Menu > Preprocessor > Modeling > Create > Lines >
Line Fillet
```

A radius RAD of $t/2 = 0.001$ m is used. As explained for the example problem in §4.10, the area enclosed by the fillet and the original corner is then generated and added to the original area. The steps are

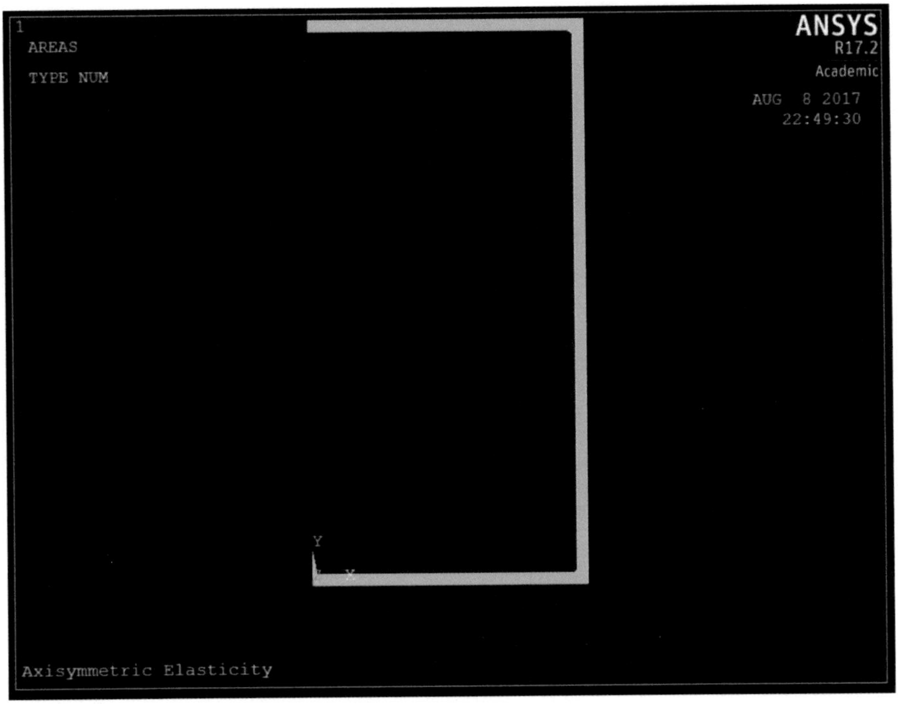

Fig. 5.4 Axisymmetric elasticity example problem: geometric modeling

```
Main Menu > Preprocessor > Modeling > Create > Areas >
Arbitrary > By Lines
```

The three lines (two straight plus the curved line at the fillet) are now selected. The resulting enclosed area is then added to the original area:

```
Main Menu > Preprocessor > Modeling > Operate > Booleans >
Add > Areas
```

The above steps are repeated for the second fillet at the other internal corner. Refer to Fig. 5.4 to see the complete geometry. The static analysis type is prescribed:

```
Main Menu > Solution > Analysis Type > New Analysis >
Static
```

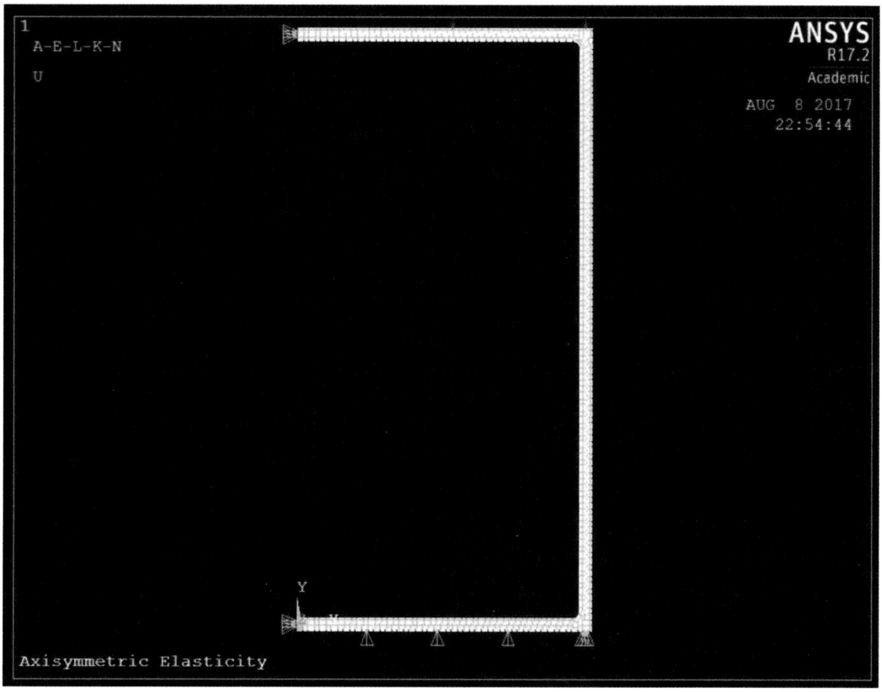

Fig. 5.5 Axisymmetric example problem: mesh and boundary conditions

Boundary conditions are now applied to the geometry. Guided by the right side of Fig. 5.2, we first impose the essential boundary conditions of null axial displacement u_y on the bottom edge of the body at $y = 0$:

```
Main Menu > Preprocessor > Define Loads > Apply >
Structural > Displacement > On Lines
```

Choose the line or lines along the bottom and constrain their y-displacement component UY to zero. Then we repeat the process for the two short vertical edges on the left side, setting radial displacement UX to zero (i.e., $u_x = 0$) on these two lines. This BC prevents the can from opening ($u_x > 0$) or collapsing ($u_x < 0$) along the centerline. The combination of the two sets of essential boundary conditions leads to a fixed keypoint at the origin that will serve to prevent rigid body translation. Natural BCs corresponding to the pressure load[1] acting on the top surface are then imposed by

[1] By default, a positive pressure boundary condition in ANSYS corresponds to compression or a negative hydrostatic stress.

```
Main Menu > Preprocessor > Define Loads > Apply >
Structural > Pressure > On Lines
```

Enter the value of $p = 2e6$ for the line corresponding to $y = h$. The mesh size is for this example is then chosen:

```
Main Menu > Preprocessor > Meshing > Size Cntrls > Manual
Size > Areas > All Areas
```

A value of 0.0004 is entered for the element edge length that will produce five elements along through the thickness t of the can/cylinder. The mesh is then generated via

```
Main Menu > Preprocessor > Mesh > Areas > Free > Pick
```

Choose the sole existing area and click "OK". The mesh is shown in Fig. 5.5 along with the symbolic boundary conditions.

5.8.3 Solution and Post-processing

The FE solution is obtained in ANSYS for the present static analysis by the usual sequence

```
Main Menu > Solution > Solve > Current LS > OK
```

In post-processing, the following items are of high interest: the deformed shape of the body, the maximum deflection over all nodes in the body, and the distributions of axial stress and scalar effective stress in the body. The deformed shape can be viewed by the command

```
Main Menu > General Postproc > Plot Results > Deformed
Shape
```

In order to see the full geometry corresponding to the 3-D structure as in Fig. 5.6, the following steps are undertaken:

```
Utility Menu > PlotCtrls > Style > Symmetry Expansion > 2D
Axi-Symmetric > Full Expansion
```

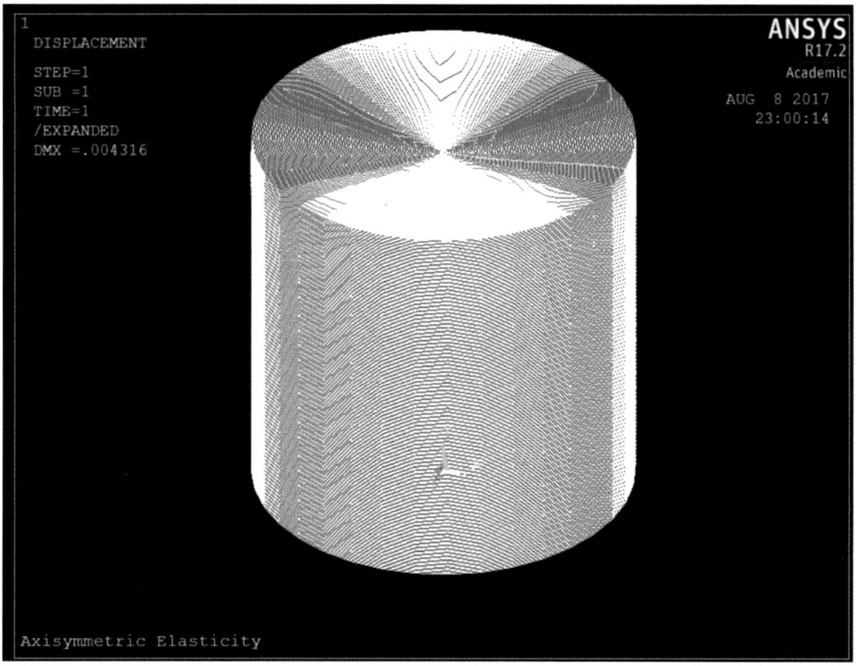

Fig. 5.6 Axisymmetric elasticity example problem: deformed structure, 3-D rendering

As is the default case, ANSYS amplifies the displacement to facilitate visualization, though here only by a factor of 1.159. The maximum deflection occurs at the center point of the top of the structure, with a magnitude of 4.316 mm $\approx 2t$, downward in conjunction with the applied pressure. To return to the planar axisymmetric view,

```
Utility Menu > PlotCtrls > Style > Symmetry Expansion > 2D
Axi-Symmetric > No Expansion
```

Next, the element solution for the axial stress σ_{yy} is plotted by the steps

```
Main Menu > General Postproc > Plot Results > Element
Solution > Y-Component of stress
```

We are interested in the average stress supported by the wall of the can. This value can be obtained by selecting five elements in series along the wall via

```
Utility Menu > Select Entities > Elements > By Num/Pick
```

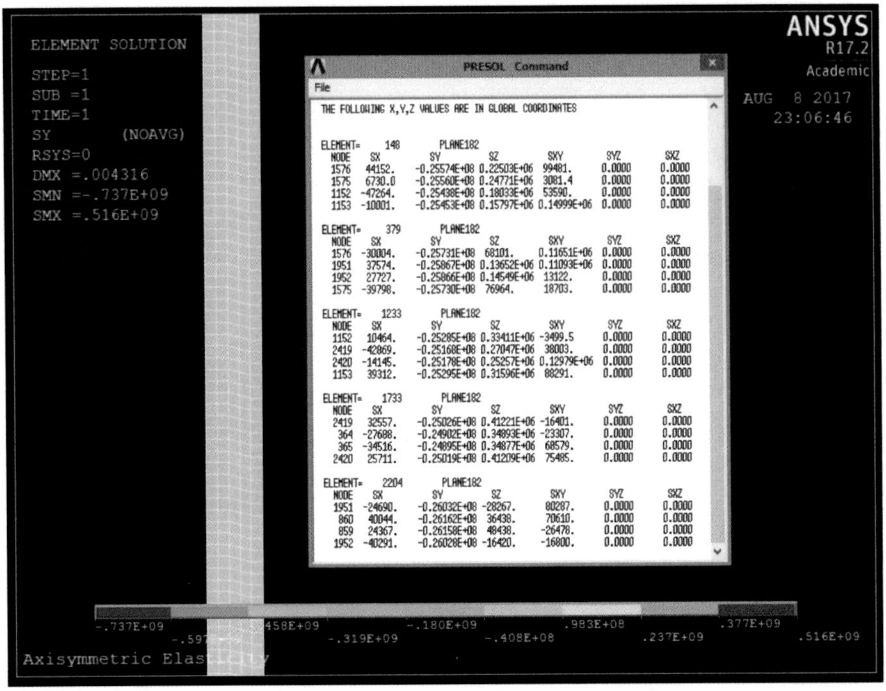

Fig. 5.7 Axisymmetric elasticity example problem: axial stress contour, zoomed view on lateral edge

The stress supported by these elements is then listed:

```
Main Menu > General Postproc > List Results > Element
Solution > Y-Component of stress
```

The contour plot and results listing for a section of the lateral wall of the structure at $y \approx h/2$ are shown in Fig. 5.7. The average stress supported by the selected elements is ≈ 25 or 26 MPa, negative in sign for compression. To validate this result, we note that the average stress S from simple statics acting on the structure in Fig. 5.2 should be the pressure p times the area of the top surface $\pi(R_0)^2$ which gives the total force F, divided by the cross-sectional area A of the lateral edge:

$$S = \frac{F}{A} = \frac{\pi(R_0)^2 p}{\pi[(R_0)^2 - (R_0 - t)^2]} = 25.51 \text{ MPa}. \tag{5.86}$$

The final quantity of interest is the scalar effective stress, i.e., the von Mises stress. As introduced in §4.10, this quantity is a non-negative scalar measure of the intensity of the stress tensor that excludes effects of hydrostatic pressure. In

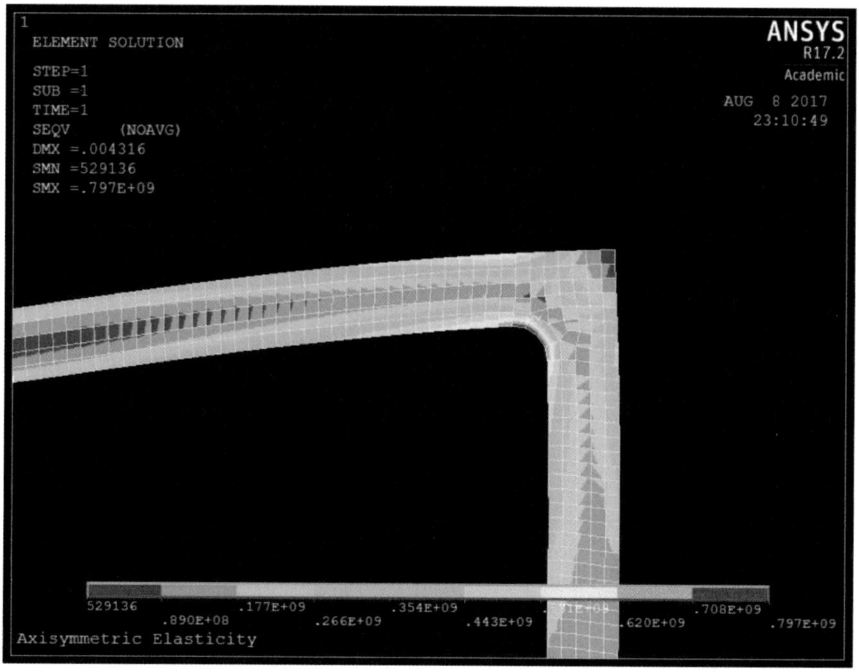

Fig. 5.8 Planar elasticity example problem: effective stress contour, zoomed view at top right corner

engineering metals, plastic yielding often occurs when the von Mises stress exceeds a threshold value. For an axisymmetric problem, the von Mises stress is computed from components of the stress tensor in the (r, θ, z) coordinate system as

$$\sigma_v = \left(\frac{1}{2}[(\sigma_{rr} - \sigma_{\theta\theta})^2 + (\sigma_{\theta\theta} - \sigma_{zz})^2 + (\sigma_{zz} - \sigma_{rr})^2 + 6(\sigma_{rz})^2]\right)^{1/2}. \quad (5.87)$$

The transformations $\sigma_{rr} \to \sigma_{xx}, \sigma_{\theta\theta} \to \sigma_{zz}, \sigma_{zz} \to \sigma_{yy}, \sigma_{rz} \to \sigma_{xy}$ are enacted to change indices in this expression to those used in ANSYS for axisymmetric problems. The maximum value of the von Mises stress over all elements (or Gauss points) is

$$\sigma_m = \max_{(e)} \sigma_v^e. \quad (5.88)$$

The von Mises stress contour reported in Fig. 5.8 is acquired by the sequence

```
Main Menu > General Postproc > List Results > Element
Solution > von Mises stress
```

The maximum value occurs at the filleted upper inner corner[2] as shown in Fig. 5.8. This value exceeds the yield stress of the aluminum material listed in Fig. 5.2, i.e., $\sigma_m \approx 800$ MPa > 276 MPa $= \sigma_y$. Thus, plastic deformation, and perhaps collapse of the structure, would be expected for this loading protocol. It is remarked that the maximum loading that could be supported by the structure for this example assumes perfect elastic behavior and perfect axisymmetry. For a real physical system, the analyst should also consider the possibility of buckling instabilities that would break symmetry and that would be expected to severely reduce the maximum load that could be safely carried by the structure.

5.9 Problems

5.9.1. Consider a thick column made of an isotropic, linear elastic solid material as shown in Fig. 5.9. The length of the column is 10 mm and the radius of the column is 2 mm. The column is clamped at the center of one end to prevent rigid body motion, and it is free to expand laterally but not axially. A distributed pressure load of magnitude $p = 1$ GPa is applied normal to the opposite face of the column, creating an effectively uniaxial compressive stress state. Specifically for this problem, the column is considered to be made of the solid metal lead with elastic modulus $E = 16$ GPa and Poisson's ratio $\nu = 0.45$.
a. Using finite element software, analyze the problem for quasi-static loading conditions invoking an axisymmetric model. Describe the steps used to create the geometry and boundary conditions. Then obtain contour plots of the deformed shape and axial stress after the load is applied.
b. Repeat the steps of part **a.** invoking a fully 3-D model (e.g., tetrahedral or hexahedral finite elements) and comment on any differences observed in the corresponding numerical results. [Hint: you may need to constrain more than one node in a small

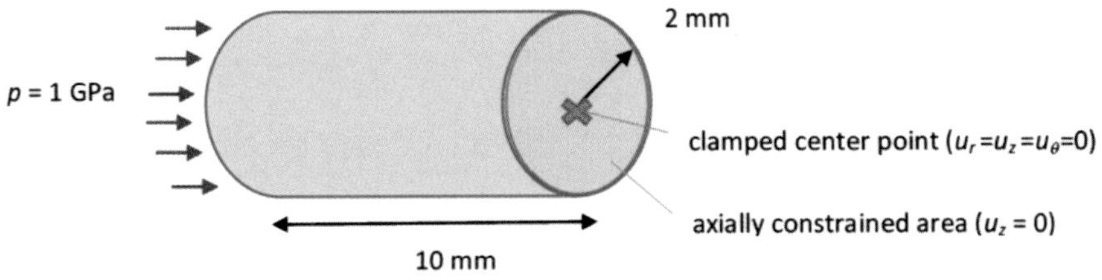

Fig. 5.9 Problem 5.9.1

[2] Had we omitted the fillet, a stress singularity would lead to a mesh dependent solution as in the example problem of §4.10.

region in the vicinity of the centerline on the axially restrained area to prevent rigid rotations about the axis of the cylinder.]

References

1. L.E. Malvern, *Introduction to the Mechanics of a Continuous Medium* (Prentice-Hall, Englewood Cliffs, NJ, 1969)
2. J.D. Clayton, *Differential Geometry and Kinematics of Continua* (World Scientific, Singapore, 2014)
3. J.D. Clayton, *Nonlinear Mechanics of Crystals* (Springer, Dordrecht, 2011)
4. C. Teodosiu, *Elastic Models of Crystal Defects* (Springer, Berlin, 1982)
5. J.D. Clayton and J.J. Rencis, Numerical integration in the axisymmetric finite element formulation, *Advances in Engineering Software* **31**, 137–141 (2000)

Chapter 6
Time Dependence

Abstract This chapter addresses the general case of time dependent problems, wherein field variables can vary with both spatial position and time. Such problems are also referred to as dynamic events, in contrast to static or quasi-static problems of equilibrium considered in prior chapters. The first part of this chapter is restricted to problems whose fields may vary with time and no more than one spatial dimension. A general partial differential equation is introduced that is representative of dynamic heat transfer and dynamic solid mechanics problems. This equation may be elliptic, parabolic, or hyperbolic depending on coefficients of differential terms. Parabolic equations representative of heat transfer are then considered, with analytical solutions derived first, followed by derivation of finite element equations for transient linear heat conduction problems. Hyperbolic equations representative of solid mechanics are considered next, again with a focus on select analytical solutions and finite element equations for linear elasto-dynamic problems. Issues important for time integration of the dynamic equations of heat conduction and physical momentum are presented. Finite element methods for fully dynamic analysis, modal analysis, and harmonic analysis are then discussed in turn.

In most real physical systems, observables vary with both time and space. Prior chapters have considered only spatial dependence, with a focus on 1-D and 2-D problems. Time dependent problems are now analyzed, first via analytical mathematical methods and then via finite element methods.

6.1 Initial Boundary Value Problems

A general partial differential equation (PDE) is now introduced that proves useful for categorizing problems whose solutions may vary in both space and time, or more generally may vary with two independent variables. This equation, containing derivatives of up to order two of the primary variable, is written as

$$A\frac{\partial^2\Phi}{\partial x^2}+B\frac{\partial^2\Phi}{\partial x\partial t}+C\frac{\partial^2\Phi}{\partial t^2}+D\frac{\partial\Phi}{\partial x}+E\frac{\partial\Phi}{\partial t}+F\Phi+G=0, \tag{6.1}$$

where A,B,C,D,E,F,G are constants and $\Phi = \Phi(x,t)$ is the primary field variable depending potentially on two independent coordinates or on two independent variables, the latter here taken as x and t. Coefficients of the second derivative terms enable classification of this PDE into one of three types:

- **Elliptic**: $B^2 - AC < 0$;
- **Parabolic**: $B^2 - AC = 0$;
- **Hyperbolic**: $B^2 - AC > 0$.

Of particular interest to governing equations of transient heat conduction and dynamic solid mechanics are parabolic and hyperbolic PDEs, respectively. These two classes of differential equations are discussed in more detail in the following two subsections.

6.1.1 Parabolic Equations

Consider the physical problem of transient heat conduction in a Fourier conductor, where the general governing equation in coordinate-free notation is

$$\rho c\frac{\partial T}{\partial t} = \nabla\cdot(\boldsymbol{k}\nabla T)+f. \tag{6.2}$$

Notation in (6.2) is as follows: T is the temperature, t is time, $\boldsymbol{k}$ is the matrix/tensor of thermal conductivity, and f accounts for heat sources. Mass density and specific heat capacity are the material properties denoted by ρ and c, respectively. In general, T, f, c, ρ, and $\boldsymbol{k}$ may all depend on position within the body, and T and f may also be functions of time. Equation (6.2) is the more general dynamic version of the governing equation for heat conduction given in static equilibrium form in (4.5) of Chapter 4. This equation corresponds thermodynamically to the local conservation of energy or the First law of Thermodynamics. The heat flux vector obeys Fourier's law:

$$\boldsymbol{q} = -\boldsymbol{k}\nabla T. \tag{6.3}$$

In one spatial dimension (x), (6.2) becomes, for a 1-D conductor with conductivity k and heat flux $q = -k(\partial T/\partial x)$,

$$\rho(x)c(x)\frac{\partial T(x,t)}{\partial t} = \frac{\partial}{\partial x}\left[k(x)\frac{\partial T(x,t)}{\partial x}\right] + f(x,t). \tag{6.4}$$

Physically, this governing equation describes time dependent heat flow through a bar or channel, for example. Categorization of transient heat conduction in the context of (6.1) requires that the conductor have constant and uniform material properties, and that the forcing function f is constant (or zero):

$$\frac{\partial^2 T}{\partial x^2} - \frac{\rho c}{k}\frac{\partial T}{\partial t} + \frac{f}{k} = 0. \tag{6.5}$$

From (6.5) with $T(x,t) \to \Phi(x,t)$, it follows that $A = 1$, $B = C = D = F = 0$, $E = -\rho c/k < 0$, and $G = f/k$. Thus, $B^2 - AC = 0$, and we have verified that this problem is of parabolic type.

It is instructive to derive the general analytical solution to the homogeneous form of (6.5), which corresponds to $f = 0$:

$$\frac{\partial^2 T}{\partial x^2} = \frac{\rho c}{k}\frac{\partial T}{\partial t}. \tag{6.6}$$

The separation of variables technique is used, where the homogeneous solution is assumed to be of the following product form:

$$T(x,t) = \varphi(x)\psi(t), \tag{6.7}$$

with φ and ψ each functions of a single independent variable as indicated by their arguments. Substituting (6.7) into (6.6) gives

$$\psi(t)\frac{\partial^2\varphi(x)}{\partial x^2} = \varphi(x)\frac{\rho c}{k}\frac{\partial\psi(t)}{\partial t}. \tag{6.8}$$

Isolating terms that depend on either x or t alone to either side of the equation, we obtain

$$\frac{1}{\psi}\frac{\partial\psi}{\partial t} = \frac{k}{\rho c}\frac{1}{\varphi}\frac{\partial^2\varphi}{\partial x^2} = -\Lambda. \tag{6.9}$$

Since terms on each side of the first equality in (6.9) are each a function of only one independent variable, each side must be equal to a constant, denoted here by (negative) Λ. It follows that (6.9) can be written as two separate homogeneous ordinary differential equations (ODEs):

$$\frac{\mathrm{d}\psi}{\mathrm{d}t} + \Lambda\psi = 0, \qquad \frac{\mathrm{d}^2\varphi}{\mathrm{d}x^2} + \left(\frac{\Lambda\rho c}{k}\right)\varphi = 0. \tag{6.10}$$

The general solution of the first order, time dependent ODE is

$$\psi(t) = \psi_0\exp(-\Lambda t). \tag{6.11}$$

Constant ψ_0 can be determined from initial conditions in conjunction with subsequent results. The general solution of the position dependent, second order ODE is

$$\varphi(x) = A_1 \sin\left(x\sqrt{\Lambda\rho c/k}\right) + A_2 \cos\left(x\sqrt{\Lambda\rho c/k}\right). \tag{6.12}$$

Constants A_1 and A_2 are determined from boundary conditions. Combining (6.11) and (6.12) and absorbing ψ_0 into function $\phi(x)$ and new constants B_1 and B_2, the total solution $T(x,t) = \varphi(x)\psi(t)$ becomes

$$\begin{aligned} T(x,t) &= \psi_0\varphi(x)\exp(-\Lambda t) = \phi(x)\exp(-\Lambda t) \\ &= \left[B_1 \sin\left(x\sqrt{\Lambda\rho c/k}\right) + B_2 \cos\left(x\sqrt{\Lambda\rho c/k}\right)\right]\exp(-\Lambda t). \end{aligned} \tag{6.13}$$

The associated eigenvalue problem is obtained by substituting the form of the general solution following the second equality in (6.13) into the original PDE for the homogeneous initial boundary value problem, (6.6), differentiating with respect to time, and then dividing through by the exponential term:

$$\frac{\mathrm{d}^2\phi(x)}{\mathrm{d}x^2} - \frac{\Lambda\rho c}{k}\phi(x) = 0. \tag{6.14}$$

Determination of eigenvalue(s) and eigenmode(s) or mode shape(s) ϕ_i that satisfy (6.14) constitutes solution of this eigenvalue problem. The finite element (FE) method will now be used to discretize the space domain. Later FEA will be used to more fully treat this problem.

In FE analysis of transient problems, shape functions are implemented for interpolation of field variables at locations distinct from nodal coordinates at any given value of time t. Nodal values of the primary field variable, e.g., temperature T in the present context, depend on time but not space for any single node. Let $T_i^{(e)}$ denote the discrete value of temperature for element e at node i with coordinate x_i. Then the time dependent temperature field is interpolated as

$$T(x,t) = N_i(x)T_i^{(e)}(t), \tag{6.15}$$

where the summation convention applies to index i that spans the number of nodes in the element. Shape functions $N_i(x)$ for 1-D discretization are identical to those introduced in §2.3.3, for example (2.112) for a linear bar element of length L with two nodes located at $x = 0, L$. Derivation of the finite element equations for dynamics proceeds with introduction of time- and space-dependent weight function $w(x,t)$, differentiable but otherwise arbitrary. This function is interpolated analogously to temperature, i.e., using the same shape functions:

$$w(x,t) = N_i(x)w_i(t), \tag{6.16}$$

with w_i depending generally on time but not position at any node i at a fixed location x_i in the FE mesh.

Derivation of the weak form equation for a single element in dynamic thermal problems proceeds as follows. The strong form PDE is (6.4), repeated below with arguments suppressed and all terms moved to the left side for the general non-homogeneous case ($f \neq 0$):

$$\rho c \frac{\partial T}{\partial t} - \frac{\partial}{\partial x}\left(k \frac{\partial T}{\partial x}\right) - f = 0. \tag{6.17}$$

This equation is next multiplied by w and integrated over the domain for a single 1-D bar element $\{\Omega^e : x \in [0, L]\}$:

$$\int_{\Omega^e} \rho c w \frac{\partial T}{\partial t} \mathrm{d}x - \int_{\Omega^e} w \frac{\partial}{\partial x}\left(k \frac{\partial T}{\partial x}\right) \mathrm{d}x - \int_{\Omega^e} w f \mathrm{d}x = 0. \tag{6.18}$$

The second term is integrated by parts:

$$\int_{\Omega^e} w \frac{\partial}{\partial x}\left(k \frac{\partial T}{\partial x}\right) \mathrm{d}x = \left(wk \frac{\partial T}{\partial x}\right)\bigg|_0^L - \int_{\Omega^e} \frac{\partial w}{\partial x} k \frac{\partial T}{\partial x} \mathrm{d}x. \tag{6.19}$$

Substituting (6.19) into (6.18) and defining the boundary heat flux contributions as $Q_1 = -k(\partial T/\partial x)|_0$ and $Q_2 = k(\partial T/\partial x)|_L$ results in

$$\int_{\Omega^e} \rho c w \frac{\partial T}{\partial t} \mathrm{d}x + \int_{\Omega^e} \frac{\partial w}{\partial x} k \frac{\partial T}{\partial x} \mathrm{d}x = w(0) Q_1 + w(L) Q_2 + \int_{\Omega^e} w f \mathrm{d}x. \tag{6.20}$$

The integro-differential equation in (6.20) is the weak form equation for a single dynamic thermal finite element.

Governing equations in vector-matrix form are obtained following substitution of the discretizations in (6.15) and (6.16) into (6.20):

$$\begin{aligned} \int_{\Omega^e} w_i \rho c N_i N_j \dot{T}_j^{(e)} \mathrm{d}x + \int_{\Omega^e} w_i \frac{\partial N_i}{\partial x} k \frac{\partial N_j}{\partial x} T_j^{(e)} \mathrm{d}x & \\ = w_i Q_i + \int_{\Omega^e} w_i N_i f \mathrm{d}x. & \end{aligned} \tag{6.21}$$

Notice that temperature rates have been interpolated via time differentiation of the sum in (6.15):

$$\frac{\partial T}{\partial t}(x,t) = N_j(x) \dot{T}_j^{(e)}(t), \qquad \dot{T}_j^{(e)}(t) = \frac{\mathrm{d}}{\mathrm{d}t}[T_j^{(e)}(t)]. \tag{6.22}$$

Presuming that the weight function w and its nodal values w_i are arbitrary, the latter can be factored and eliminated from (6.21) to give

$$\int_{\Omega^e} \rho c N_i N_j \dot{T}_j^{(e)} \mathrm{d}x + \int_{\Omega^e} \frac{\partial N_i}{\partial x} k \frac{\partial N_j}{\partial x} T_j^{(e)} \mathrm{d}x = Q_i + \int_{\Omega^e} N_i f \mathrm{d}x. \tag{6.23}$$

This set of i scalar equations (with $i = 1,2$ for a linear element) can be written in vector-matrix form as

$$[\boldsymbol{M}^{(e)}]\{\dot{\boldsymbol{T}}\} + [\boldsymbol{k}^{(e)}]\{\boldsymbol{T}\} = \{\boldsymbol{Q}^{(e)}\} + \{\boldsymbol{f}^{(e)}\} = \{\boldsymbol{F}^{(e)}\}, \tag{6.24}$$

where $\dot{\boldsymbol{T}}$ and $\boldsymbol{T}$ are column vectors of nodal temperature rates and nodal temperatures, respectively, that can be factored outside the integrals. The usual element stiffness matrix and B-matrix of shape function derivatives are

$$[\boldsymbol{k}^{(e)}] = \int_0^L [\boldsymbol{B}]^{\mathrm{T}} k [\boldsymbol{B}] \mathrm{d}x, \qquad [\boldsymbol{B}] = \frac{\mathrm{d}}{\mathrm{d}x}[\boldsymbol{N}]. \tag{6.25}$$

Components of the boundary flux vector and point source vectors are, respectively,

$$Q_i^{(e)} = Q_i, \qquad f_i^{(e)} = \int_0^L N_i f \mathrm{d}x. \tag{6.26}$$

The sum of the two vectors in (6.26) is the total force vector $\boldsymbol{F}^{(e)}$ on the far right in (6.24). Finally, the mass matrix for 1-D heat conduction problems is defined as the integral

$$[\boldsymbol{M}^{(e)}] = \int_{\Omega^e} \rho c [\boldsymbol{N}]^{\mathrm{T}} [\boldsymbol{N}] \mathrm{d}x \quad \leftrightarrow \quad M_{ij}^{(e)} = \int_0^L \rho(x) c(x) N_i(x) N_j(x) \mathrm{d}x. \tag{6.27}$$

The mass matrix will be considered in more detail in the context of time integration later in §6.2. The governing equation in (6.24) is called a semi-discrete equation for transient heat transfer problems or analogous parabolic problems.

6.1.2 Hyperbolic Equations

Now consider the physical problem of momentum conservation in a deformable solid body, where the general local governing equation in coordinate-free notation is [1, 2]

$$\nabla \cdot \boldsymbol{\sigma} + \boldsymbol{f} = \rho \boldsymbol{a}. \tag{6.28}$$

Here, $\boldsymbol{\sigma}$ is the symmetric stress tensor, $\boldsymbol{f}$ is the body force vector, ρ is the mass density, and $\boldsymbol{a}$ is the acceleration vector. In general, $\boldsymbol{\sigma}, \boldsymbol{f}, \rho$, and $\boldsymbol{a}$ may all depend on position within the body, and $\boldsymbol{\sigma}$, $\boldsymbol{a}$, and $\boldsymbol{f}$ may also be functions of time. Equation (6.28) is the more general dynamic version of the governing equation for local (static) equilibrium in solid mechanics introduced in (4.54) of Chapter 4. The right hand side of (6.28) accounts for inertia. In one spatial dimension (e.g., x), (6.28) becomes, for a 1-D linear elastic body with elastic modulus E and stress-strain constitutive law $\sigma = E\varepsilon = E(\partial u/\partial x)$,

$$\rho(x)\frac{\partial^2 u(x,t)}{\partial t^2} = \frac{\partial}{\partial x}\left[E(x)\frac{\partial u(x,t)}{\partial x}\right] + f(x,t). \tag{6.29}$$

The displacement function is $u(x,t)$. Physically, this governing equation describes time dependent axial deformation and motion of an elastic bar, for example. Categorization of solid dynamics in the context of (6.1) requires that the body have constant material properties, and that the body force per unit volume f is constant (or zero), leading to

$$\frac{\partial^2 u}{\partial x^2} - \frac{\rho}{E}\frac{\partial^2 u}{\partial t^2} + \frac{f}{E} = 0. \tag{6.30}$$

From (6.30) with $u(x,t) \to \Phi(x,t)$, it follows that $A = 1$, $C = -\rho/E < 0$, $B = D = E = F = 0$, and $G = f/E$, Duplicate use of notation E is invoked for the elastic modulus and the fifth coefficient in (6.1), but context here should alleviate the possibility of confusion. Thus, $B^2 - AC = \rho/E > 0$, and we have verified that the problem is of hyperbolic type.

Following the same pattern of discussion used in §6.1.1, we now derive the general analytical solution to the following homogeneous form ($f = 0$) of conservation law (6.30):

$$\frac{\partial^2 u}{\partial x^2} = \frac{\rho}{E}\frac{\partial^2 u}{\partial t^2}. \tag{6.31}$$

The separation of variables technique is again used, where the homogeneous solution is posited to be of the general product form

$$u(x,t) = \varphi(x)\psi(t), \tag{6.32}$$

with φ and ψ each functions of a single independent variable as indicated by their arguments. Substituting (6.32) into (6.31) produces

$$\psi(t)\frac{\partial^2 \varphi(x)}{\partial x^2} = \varphi(x)\frac{\rho}{E}\frac{\partial^2 \psi(t)}{\partial t^2}. \tag{6.33}$$

Isolating terms that depend on either x or t alone to each side of the equation, we arrive at

$$\frac{1}{\psi}\frac{\partial^2 \psi}{\partial t^2} = \frac{E}{\rho}\frac{1}{\varphi}\frac{\partial^2 \varphi}{\partial x^2} = -\omega^2. \tag{6.34}$$

Following prior arguments of §6.1.1, groups on each side of the first equality must be equal to a constant, denoted here by (negative) ω^2. Thus (6.34) can be written as two separate homogeneous ordinary differential equations:

$$\frac{\mathrm{d}^2\psi}{\mathrm{d}t^2} + \omega^2\psi = 0, \qquad \frac{\mathrm{d}^2\varphi}{\mathrm{d}x^2} + \left(\frac{\omega^2\rho}{E}\right)\varphi = 0. \tag{6.35}$$

The general solution of the second-order time dependent ODE is

$$\psi(t) = C_1\cos(\omega t) + C_2\sin(\omega t), \tag{6.36}$$

where C_1 and C_2 are constants that depend on initial and boundary conditions. The general solution of the position dependent ODE is

$$\varphi(x) = A_1 \sin\left(\omega x\sqrt{\rho/E}\right) + A_2 \cos\left(\omega x\sqrt{\rho/E}\right). \tag{6.37}$$

Integration constants A_1 and A_2 are typically determined from boundary conditions. Substituting (6.36) into (6.32), the total solution $u(x,t) = \varphi(x)\psi(t)$ can be written as

$$u(x,t) = \varphi(x)\left[C_1 \cos(\omega t) + C_2 \sin(\omega t)\right] = \phi(x)\exp(-i\omega t). \tag{6.38}$$

Following the second equality, $i = \sqrt{-1}$ and function $\phi(x)$ absorbs the products of constant coefficients contained in $\varphi(x)$ and $\psi(t)$. The associated eigenvalue problem is obtained by substituting the form of the general solution following the second equality in (6.38) into the original PDE for the homogeneous initial boundary value problem, (6.31), taking the associated two time derivatives, and then dividing through by the exponential term:

$$\frac{\mathrm{d}^2\phi(x)}{\mathrm{d}x^2} - \frac{\omega^2\rho}{E}\phi(x) = 0. \tag{6.39}$$

Determination of eigenvalue(s) and eigenmode(s) or mode shape(s) ϕ_i that satisfy (6.39) constitutes solution of this eigenvalue problem for solid dynamics. Notice that (6.39) becomes identical to the eigenvalue equation for heat conduction in (6.14) if we let $\omega^2\rho/E \rightarrow \Lambda\rho c/k$. The finite element method will now be used to discretize the space domain and then later in Chapter 6 will be used to revisit this eigenvalue problem.

In FE analysis of dynamic solid mechanics problems, just as in transient thermal problems, shape functions are invoked for representation of field variables at locations distinct from nodal coordinates at any time t. Nodal values of the primary field variable, displacement u in the present context, depend on time but not space for any single node. Let $u_i^{(e)}$ denote the discrete value of displacement for element e at node i with initial coordinate x_i. Then the time dependent displacement field is interpolated as

$$u(x,t) = N_i(x)u_i^{(e)}(t), \tag{6.40}$$

where the summation convention applies to index i that spans the number of nodes in the element. Shape functions $N_i(x)$ for 1-D discretization are identical to those introduced in §2.3.3, for example (2.112) for a linear bar element covering the domain $\{\Omega^e : x \in [0,L]\}$. Derivation of the finite element equations for solid dynamics requires time- and space-dependent weight function $w(x,t)$, differentiable but otherwise arbitrary. This function is interpolated analogously to displacement, with identical shape functions $N_i(x)$:

$$w(x,t) = N_i(x)w_i(t), \tag{6.41}$$

with w_i depending generally on time but not position at any node i at fixed (initial) coordinate x_i.

Derivation of the weak form equation for a single element in solid elastodynamics problems proceeds as follows, paralleling the derivation for transient heat conduction in §6.1.1. The strong form PDE for 1-D dynamics of an elastic bar is (6.29), repeated below with functional arguments suppressed and all terms moved to the left side:

$$\rho \frac{\partial^2 u}{\partial t^2} - \frac{\partial}{\partial x}\left(E\frac{\partial u}{\partial x}\right) - f = 0. \tag{6.42}$$

This equation is then multiplied by w and integrated over the domain for a single 1-D bar element:

$$\int_{\Omega^e} \rho w \frac{\partial^2 u}{\partial t^2}\mathrm{d}x - \int_{\Omega^e} w \frac{\partial}{\partial x}\left(E\frac{\partial u}{\partial x}\right)\mathrm{d}x - \int_{\Omega^e} w f \mathrm{d}x = 0. \tag{6.43}$$

The second term on the left is integrated by parts:

$$\int_{\Omega^e} w \frac{\partial}{\partial x}\left(E\frac{\partial u}{\partial x}\right)\mathrm{d}x = \left(wE\frac{\partial u}{\partial x}\right)\Big|_0^L - \int_{\Omega^e} \frac{\partial w}{\partial x} E \frac{\partial u}{\partial x}\mathrm{d}x. \tag{6.44}$$

Substituting (6.44) into (6.43) and defining the boundary traction contributions as $Q_1 = -E(\partial u/\partial x)|_0 = \sigma_0$ and $Q_2 = E(\partial u/\partial x)|_L = \sigma_L$ produces the following equation:

$$\int_{\Omega^e} \rho w \frac{\partial^2 u}{\partial t^2}\mathrm{d}x + \int_{\Omega^e} \frac{\partial w}{\partial x} E \frac{\partial u}{\partial x}\mathrm{d}x = w(0)Q_1 + w(L)Q_2 + \int_{\Omega^e} w f \mathrm{d}x. \tag{6.45}$$

The integro-differential equation in (6.45) is the weak form equation for a single dynamic finite element in 1-D linear elasticity.

Governing equations in vector-matrix form are obtained following substitution into (6.45) of the expansions in terms of shape functions N_i or N_j in (6.40) and (6.41), where subscripts i or j cover all nodal degrees of freedom, two for the case of a linear bar element:

$$\begin{aligned}\int_{\Omega^e} w_i \rho N_i N_j \ddot{u}_j^{(e)}\mathrm{d}x + \int_{\Omega^e} w_i \frac{\partial N_i}{\partial x} E \frac{\partial N_j}{\partial x} u_j^{(e)}\mathrm{d}x \\ = w_i Q_i + \int_{\Omega^e} w_i N_i f \mathrm{d}x.\end{aligned} \tag{6.46}$$

Accelerations are interpolated via twice time differentiation of (6.40), recalling that shape functions do not depend on time:

$$\frac{\partial^2 u}{\partial t^2}(x,t) = N_j(x)\ddot{u}_j^{(e)}(t), \qquad \ddot{u}_j^{(e)}(t) = \frac{\mathrm{d}^2}{\mathrm{d}t^2}[u_j^{(e)}(t)]. \tag{6.47}$$

Taking advantage of arbitrariness of weight function w and its nodal values w_i, the latter can be factored and eliminated from (6.46) to result in

$$\int_{\Omega^e} \rho N_i N_j \ddot{u}_j^{(e)} \mathrm{d}x + \int_{\Omega^e} \frac{\partial N_i}{\partial x} E \frac{\partial N_j}{\partial x} u_j^{(e)} \mathrm{d}x = Q_i + \int_{\Omega^e} N_i f \mathrm{d}x. \tag{6.48}$$

This set of i scalar momentum conservation equations (with $i = 1,2$ for a linear element) can be written concisely in vector-matrix form as

$$[\boldsymbol{M}^{(e)}]\{\ddot{\boldsymbol{u}}\} + [\boldsymbol{k}^{(e)}]\{\boldsymbol{u}\} = \{\boldsymbol{Q}^{(e)}\} + \{\boldsymbol{f}^{(e)}\} = \{\boldsymbol{F}^{(e)}\}, \tag{6.49}$$

where $\ddot{\boldsymbol{u}}$ and $\boldsymbol{u}$ are respective column vectors of nodal accelerations and nodal displacements, both of which can be factored outside the corresponding integrals. The usual element stiffness matrix and B-matrix for an elastic bar are, respectively,

$$[\boldsymbol{k}^{(e)}] = \int_0^L [\boldsymbol{B}]^{\mathrm{T}} E [\boldsymbol{B}] \mathrm{d}x, \qquad [\boldsymbol{B}] = \frac{\mathrm{d}}{\mathrm{d}x}[\boldsymbol{N}]. \tag{6.50}$$

Components of the boundary traction or stress vector and body force vectors are respectively defined as

$$Q_i^{(e)} = Q_i, \qquad f_i^{(e)} = \int_0^L N_i f \mathrm{d}x. \tag{6.51}$$

The sum of the two vectors to the right of the first equality in (6.49) is the total mechanical load vector $\boldsymbol{F}^{(e)}$. Finally, the mass matrix for 1-D dynamic solid mechanics problems is defined as the following integral over the element:

$$[\boldsymbol{M}^{(e)}] = \int_{\Omega^e} \rho [\boldsymbol{N}]^{\mathrm{T}} [\boldsymbol{N}] \mathrm{d}x \quad \leftrightarrow \quad M_{ij}^{(e)} = \int_0^L \rho(x) N_i(x) N_j(x) \mathrm{d}x. \tag{6.52}$$

This mass matrix, which differs from that of heat transfer in (6.27) only by its lack of incorporation of heat capacity c in its integrand, will be considered in more detail in the context of time integration later in §6.2. The governing equation in (6.49) is a semi-discrete equation for transient/dynamic solid mechanics of elastic bodies, or for analogous hyperbolic problems.

6.2 Time Integration

Governing dynamic equations such as (6.24) and (6.49) are called semi-discrete because, although these equations are discrete in space, they are continuous in time. Numerical integration of an equation with respect to time involves determination of the solution at distinct time increments, i.e., time discretization.

Let $U(x,t)$ denote an arbitrary primary solution variable, which could be temperature T or displacement u in the context of §6.1.1 or §6.1.2, respectively. Define a fixed time increment as the finite difference

$$\Delta t = t_n - t_{n-1} = t_{n+1} - t_n, \tag{6.53}$$

where subscripts are used to label discrete values of time at successive increments. Approximations such as $\dot{U} = \Delta U/\Delta t$ are used in numerical integration schemes to compute time derivatives.

More specifically, time integration algorithms start from a point in time whose solution U_0 is known. A recursive relationship is then followed, whereby given $U_n = U(t_n)$, the immediate objective is determination of the solution at the next increment, $U_{n+1} = U(t_n + \Delta t)$.

Consider now a first-order ODE of the form in parabolic equation (6.24) repeated below with $T \to U$:

$$[\boldsymbol{M}]\{\dot{\boldsymbol{U}}\} + [\boldsymbol{K}]\{\boldsymbol{U}\} = \{\boldsymbol{F}\}. \tag{6.54}$$

Here, $\boldsymbol{M}$ and $\boldsymbol{K}$ represent mass and stiffness matrices, respectively, for the assembled and possibly condensed global system of dynamic equations. The vector of generalized nodal degrees of freedom for the global system is $\boldsymbol{U}$. The generalized alpha family of algorithms for integration of this type of ODE can be cast as the vector-matrix equations

$$[\boldsymbol{M}]\{(\boldsymbol{U}_{n+1} - \boldsymbol{U}_n)/\Delta t\} + [\boldsymbol{K}]\{\alpha \boldsymbol{U}_{n+1} + (1-\alpha)\boldsymbol{U}_n\} = \{\alpha \boldsymbol{F}_{n+1} + (1-\alpha)\boldsymbol{F}_n\}, \tag{6.55}$$

where the scalar parameter $\alpha \in [0,1]$ is prescribed by the analyst. When $\alpha = 0$, the integration scheme is fully explicit and corresponds to a forward Euler or forward difference approximation. When $\alpha = 1$, the integration scheme is fully implicit and corresponds to a backward Euler or backward difference approximation. The so-called trapezoidal rule or midpoint rule is obtained for $\alpha = \frac{1}{2}$. The alpha family applies for parabolic equations, i.e., those differential equations that are first order in time. Similar, but necessarily more involved, algorithms such as the Newmark family (which incorporates two parameters) are available for time integration of second-order differential equations, e.g., those of hyperbolic type characteristic of solid mechanics initial boundary value problems.

By multiplying both sides of (6.55) by Δt and applying rules of linear algebra, the vector value of the primary solution variable at time t_{n+1} can be isolated on one side of the equation as follows:

$$\begin{aligned}\{\boldsymbol{U}_{n+1}\} = ([\boldsymbol{M}] + \alpha\Delta t[\boldsymbol{K}])^{-1}\,&([\boldsymbol{M}] - (1-\alpha)\Delta t[\boldsymbol{K}])\{\boldsymbol{U}_n\} \\ &+ \{\alpha \boldsymbol{F}_{n+1} + (1-\alpha)\boldsymbol{F}_n\}\Delta t.\end{aligned} \tag{6.56}$$

At time t_n, the solution $\boldsymbol{U}_{n+1}$ is unknown, but all terms on the right side of (6.56) are known from the solution from the previous time increment (e.g., $\boldsymbol{U}_n$) or are prescribed (e.g., $\boldsymbol{F}_{n+1}$). Therefore, (6.56) can be used to compute the solution at time t_{n+1}.

As a convenient approximation for explicit integration schemes, the time increment for integration Δt can be chosen small enough such that $\boldsymbol{M} + \alpha\Delta t\boldsymbol{K} \approx \boldsymbol{M}$, in which case (6.56) reduces to the approximation

$$\{\boldsymbol{U}_{n+1}\} \approx \{\boldsymbol{U}_n\} - (1-\alpha)\Delta t[\boldsymbol{M}]^{-1}[\boldsymbol{K}]\,\{\boldsymbol{U}_n\} + [\boldsymbol{M}]^{-1}\{\alpha \boldsymbol{F}_{n+1} + (1-\alpha)\boldsymbol{F}_n\}\Delta t. \tag{6.57}$$

The primary numerical expense of (6.57) is evaluation of the inverse of the mass matrix. For this reason, the so-called lumped mass approximation is often invoked, whereby the original matrix $\boldsymbol{M}$ is replaced by a diagonal matrix with the diagonal entry of each row set equal to the sum of the entries of the original matrix $\boldsymbol{M}$ on that row. In the lumped mass approximation, $M_{ij}^{-1} = 1/M_{ij}$ such that the computation in (6.57) can be performed relatively quickly compared to a calculation requiring inversion of a fully populated mass matrix.

Selection of a particular time integration algorithm usually follows from consideration of accuracy, stability, and numerical efficiency (i.e., computational expense). Accuracy refers to the closeness of the integrated solution to the exact solution. Stability refers to boundedness of the solution. Stability does not necessarily ensure accuracy, and vice-versa. In general, smaller time steps improve accuracy and stability, at an increase in computational expense or a decrease in efficiency. Integration parameters such as α in (6.56) must be chosen intelligently to provide an optimum combination of accuracy, stability, and efficiency for solution of a given initial boundary value problem.

In the context of (6.55)–(6.57), and often in even more general cases, fully explicit methods tend to be computationally economical, but accuracy is only first order in Δt, and stability is conditional on the requirement that Δt be very small. For example, the Courant condition limits the maximum time step to be less than or equal to the time required for a signal to traverse the distance equal to the minimum of any edge length of an element in a FE mesh, $L_{\min}$:

$$(\Delta t)_{\max} \leq L_{\min}/C_s. \tag{6.58}$$

The velocity of the fastest important physical signal in the problem is denoted by C_s. For example, for 1-D linear isotropic elasticity, $C_s = \sqrt{E/\rho}$ is the elastic wave speed[1]. The Courant condition necessarily restricts integration of FE problems with very fine meshes (small elements) to very small time steps, resulting in possibly immense computational cost.

Fully implicit methods are more expensive per iteration than explicit methods, but stability is unconditional with respect to time step size. However, if relatively large steps Δt are used, then approximation (6.57) cannot be safely invoked, and the implicit solution suffers from the requirement of inversion of stiffness matrix $\boldsymbol{K}$, which is numerically expensive. The same remarks regarding (high) cost per iteration and (better, unconditional) stability hold for $\alpha = \frac{1}{2}$ relative to explicit integration, but accuracy is improved to second order with respect to time increment Δt.

[1] For 3-D linear elasticity, the longitudinal elastic wave speed should be used: $C_s = \sqrt{C_{11}/\rho}$ where the longitudinal modulus for an isotropic solid is $C_{11} = \lambda + 2G$ and elastic constants λ and G are defined in terms of Young's modulus E and Poisson's ratio ν in (5.62).

6.3 Finite Element Analysis

Discussion of solutions of general time dependent problems via FEA is resumed. The present section focuses on elasto-dynamics of solid bodies. However, much of the forthcoming discussion applies, with a few modifications, to transient problems in heat conduction as well as other physics or engineering problems described by similar types of PDEs.

6.3.1 Fully Dynamic Analysis

Consider the solution of continuum solid mechanics problems involving time and three space dimensions, where Cartesian coordinates $(x_1,x_2,x_3) = (x,y,z)$ are used to describe the initial position of each material point in domain Ω. The key local governing equation, specifically a conservation law for linear momentum, is in coordinate-free↔index notation:

$$\nabla \cdot \boldsymbol{\sigma} + \boldsymbol{f} = \rho \boldsymbol{a} \quad \leftrightarrow \quad \frac{\partial \sigma_{ij}}{\partial x_j} + f_i = \rho \frac{\partial^2 u_i}{\partial t^2}, \tag{6.59}$$

where $\boldsymbol{\sigma}$ is the symmetric stress tensor, $\boldsymbol{f}$ is the body force vector per unit volume, ρ is the mass density, and $\boldsymbol{a}$ is the acceleration vector. Strain-displacement relations that are ensured by differential compatibility constraints on the symmetric strain tensor $\boldsymbol{\varepsilon}$ are

$$\boldsymbol{\varepsilon} = \frac{1}{2}\left[\nabla \boldsymbol{u} + (\nabla \boldsymbol{u})^{\mathrm{T}}\right] \quad \leftrightarrow \quad \varepsilon_{ij} = \frac{1}{2}\left[\frac{\partial u_i}{\partial x_j} + \frac{\partial u_j}{\partial x_i}\right]. \tag{6.60}$$

Stress-strain relations of linear elasticity are

$$\boldsymbol{\sigma} = \boldsymbol{C} : \boldsymbol{\varepsilon} \quad \leftrightarrow \quad \sigma_{ij} = C_{ijkl}\varepsilon_{kl}. \tag{6.61}$$

Relationships in (4.60) of Chapter 4, §4.7, also apply regarding strain energy density W and symmetry of the elasticity tensor $\boldsymbol{C}$, which contains a maximum of 21 independent entries.

A set of three PDEs depending on x_i ($i = 1,2,3$) and time t with u_i the primary solution variable can be obtained by substituting (6.60) into (6.61), and then substituting this result into (6.59). These equations will contain second derivatives of displacement with respect to time and all three space coordinates. Such governing equations, which are the strong form differential equations for 3-D elasto-dynamics, can be presented in a concise form when the linear elastic material is isotropic and homogeneous, such that C_{ijkl} is independent of position x_i and is a function of two independent elastic constants λ and G:

$$C_{ijkl} = \lambda \delta_{ij}\delta_{kl} + G(\delta_{ik}\delta_{jl} + \delta_{il}\delta_{jk}). \tag{6.62}$$

Recall that δ_{ij} is the Kronecker delta: $\delta_{ij} = 1 \forall i = j$ and $\delta_{ij} = 0 \forall i \neq j$. In this case of linear isotropic elasticity in Cartesian coordinates, stress and strain density energy obey the constitutive relations

$$\sigma_{ij} = \lambda \delta_{ij} \varepsilon_{kk} + 2G\varepsilon_{ij}, \qquad W = \frac{1}{2}\lambda \varepsilon_{ii}\varepsilon_{jj} + G\varepsilon_{ij}\varepsilon_{ij}. \tag{6.63}$$

Relationships among elastic constants (λ, G) and (E, ν) are given in (5.62); other relationships among various isotropic elastic constants are readily available from references [1, 2]. Substituting from the first equality in (6.63), the local balance of linear momentum in (6.59) can be written as follows [1]:

$$G\nabla^2 \boldsymbol{u} + (\lambda + G)\nabla(\nabla \cdot \boldsymbol{u}) + \boldsymbol{f} = \rho \boldsymbol{a} \quad \leftrightarrow$$
$$G\frac{\partial^2 u_i}{\partial x_j \partial x_j} + (\lambda + G)\frac{\partial^2 u_j}{\partial x_j \partial x_i} + f_i = \rho \frac{\partial^2 u_i}{\partial t^2}. \tag{6.64}$$

Regardless of the symmetry of the material (e.g., isotropic or anisotropic elasticity tensor $\boldsymbol{C}$), the weak form equations and then the finite element equations can be derived following a procedure analogous to the 1-D case considered in §6.1.2. The end result for a single finite element is the same vector-matrix form of equation as (6.49), repeated below:

$$[\boldsymbol{M}^{(e)}]\{\ddot{\boldsymbol{u}}\} + [\boldsymbol{k}^{(e)}]\{\boldsymbol{u}\} = \{\boldsymbol{Q}^{(e)}\} + \{\boldsymbol{f}^{(e)}\} = \{\boldsymbol{F}^{(e)}\}. \tag{6.65}$$

Here, $\ddot{\boldsymbol{u}}$ and $\boldsymbol{u}$ are column vectors of nodal accelerations and displacements, each of length $3n$, where n is the number of nodes in the element. Dimensions of element mass matrix $\boldsymbol{M}^{(e)}$ and element stiffness matrix $\boldsymbol{k}^{(e)}$ are both $3n \times 3n$, and formulae analogous to (6.52) and (6.50) are used to respectively compute each of these matrices. Similarly, load vectors accounting for traction boundary conditions and body forces are of size $3n$ with components computed via 3-D analogs of those defined in (6.51).

Solution of a dynamic problem using modern FE software proceeds according to the following steps:

1. Build the geometry;
2. Construct the FE mesh (i.e., space discretization) and assemble global stiffness and mass matrices;
3. Specify initial conditions on nodal positions and nodal velocities;
4. Set solution controls (e.g., time step size Δt);
5. Apply boundary conditions;
6. Condense the global system of equations;
7. Integrate the global system for the updated nodal solution at the end of the current time increment;
8. Repeat steps 5, 6, and 7 for each time increment until end time is reached;

9. Post-process the results for desired time instances in the analysis.

Besides very simple cases such as those analyzed by the separation of variables method in §6.1.2, dynamic solid mechanics problems cannot generally be solved analytically, especially those involving complex 3-D geometries. Fortunately, those steps that are often analytically intractable, specifically steps 2 and 5–9 listed above, are performed efficiently for the analyst by any complete FE software. Step 7 is executed via procedures discussed in §6.2. With regard to post-processing, modern FE code packages enable generation of animations of deformed shapes and contours of primary and secondary variables as the analysis proceeds in time. Such animations can be quite useful for studying deformation histories and stress wave propagation in a solid.

The above discussion, and nearly all derivations for solid mechanics problems presented in this textbook, have considered the relatively simple constitutive model of linear elasticity. Modern FE software enables solution of static and dynamic problems involving much more sophisticated constitutive models, e.g., those accounting for nonlinear elasticity (finite elastic strains and/or nonlinear elastic stress-strain behavior), plasticity, fracture, and thermal strain. An example is the treatment of reference [3] that includes all of the above physical effects in conjunction with heat conduction, i.e., a multi-physics problem.

6.3.2 Modal Analysis

Modal analysis considers the eigenvalue problem in semi-discrete form, here with a focus on hyperbolic differential equations and solid mechanics applications. We consider the following governing equation of solid dynamics:

$$[\boldsymbol{M}]\{\ddot{\boldsymbol{U}}\} + [\boldsymbol{K}]\{\boldsymbol{U}\} = \{0\}. \tag{6.66}$$

Here, $\boldsymbol{M}$ is the assembled global mass matrix for the system, $\boldsymbol{U}$ is the column vector of nodal degrees of freedom (i.e., displacements), $\ddot{\boldsymbol{U}}$ is the column vector of nodal accelerations, and the right hand side is zero, corresponding to null body forces and zero applied boundary traction. Equation (6.66) is the global form of the 1-D equation of motion for a single element addressed in (6.49) of §6.1.2, restricted to the homogeneous case (i.e., null right hand side). The same general form applies for 2-D and 3-D problems in dynamic elasticity of solid bodies. Recall from (6.38) of §6.1.2 that the general solution to this homogeneous second-order PDE is of the form

$$\{\boldsymbol{U}(t)\} = \{\boldsymbol{\phi}\}\exp(-i\omega t), \tag{6.67}$$

where $\boldsymbol{\phi}$ is an eigenvector and ω is a corresponding constant, with physical units of inverse time or frequency, to be determined via solution of the eigenvalue problem.

Substituting (6.67) into (6.66) gives

$$\left(-\omega^2[\boldsymbol{M}]\{\boldsymbol{\phi}\}+[\boldsymbol{K}]\{\boldsymbol{\phi}\}\right)\exp(-i\omega t)=\{0\}. \tag{6.68}$$

Dividing through by the exponential term, we arrive at the classical eigenvalue problem

$$\left([\boldsymbol{K}]-\omega^2[\boldsymbol{M}]\right)\{\boldsymbol{\phi}\}=\{0\}. \tag{6.69}$$

If $\boldsymbol{K}$ and $\boldsymbol{M}$ are of size $n \times n$, with n the number of nodal degrees of freedom in the whole meshed domain, then the system of equations yields up to n unique eigenvalues $\lambda_i = (\omega_i)^2$, each with a corresponding unique eigenvector $\boldsymbol{\phi}_i$. In order for the system in (6.69) to possess non-trivial solution(s), the determinant of the term in parentheses must vanish, i.e.,

$$\det\left([\boldsymbol{K}]-\omega^2[\boldsymbol{M}]\right)=0, \tag{6.70}$$

a condition that yields a characteristic equation in the form of an n-th order polynomial whose roots are the eigenvalues λ_i, where $i = 1,2,\ldots n$. Once all of the modes, i.e., all of the eigenvalues and eigenvectors, have been computed, then a solution may be constructed at any time t through a linear combination:

$$\{\boldsymbol{U}(t)\}=\sum_{i=1}^{n}\{\boldsymbol{\phi}_i\}\Lambda_i(t). \tag{6.71}$$

This expression can be substituted back into the original PDE for the initial boundary value problem to produce a new system of ODEs in terms of the time dependent set of coefficients $\Lambda_i(t)$. Particular values of the latter depend on initial conditions for a given problem. For a certain set of initial conditions, a mechanical system will vibrate at one of its natural frequencies ω_i with a shape that is a scalar multiple of the corresponding mode shape $\boldsymbol{\phi}_i$. Under general initial conditions, such a system will vibrate with a combination of many frequencies and corresponding mode shapes. In such general cases, behavior of the system can be studied via the superposition of modes in (6.71). Regardless, the natural frequencies ω_i and corresponding mode shapes are the fundamental building blocks for the total solution, as illustrated conceptually in Fig. 6.1 for vibration of an elastic column.

In FEA of modal problems, the geometry and mesh are constructed in the same manner as for static problems. The only boundary conditions that are admissible are either fixed essential boundary conditions (i.e., null displacements on part of the boundary $\partial\Omega_u$) and null natural boundary conditions (i.e., null traction components on part of the boundary $\partial\Omega_t$). Nonzero applied traction, nonzero applied point forces, and any nonzero applied displacements are all mathematically inconsistent with the homogeneous form of the governing equations in (6.66) and thus are not admissible in modal analysis. Constitutive behavior is restricted to linear elasticity since superposition principles are invoked. The FE software is used to solve the eigenvalue problem, thereby computing the natural frequencies and mode shapes, which can subsequently be studied individually or in combination via post-

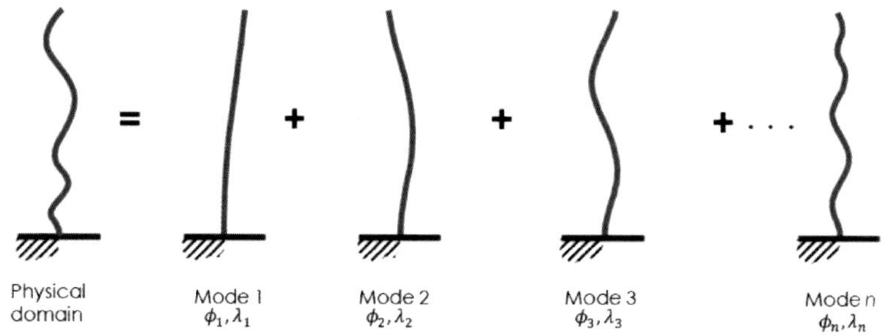

Fig. 6.1 Construction of a total solution to a modal analysis problem via superposition of n discrete modes of vibration with eigenmodes ϕ_i and corresponding eigenvalues λ_i where $i = 1, 2, \ldots n$ [image adapted from `www.svibs.com`]

processing. More specific solver and post-processing options depend on the particular FE software package employed.

6.3.3 Harmonic Analysis

Harmonic analysis is similar to modal analysis, the fundamental difference being that in the former, harmonic or sinusoidal loads are applied to the structure, rather than null external forcing functions for the case of modal analysis. The governing equation for harmonic initial boundary value problems in solid mechanics is of the form

$$[\boldsymbol{M}]\{\ddot{\boldsymbol{U}}\} + [\boldsymbol{K}]\{\boldsymbol{U}\} = \{\boldsymbol{F}\} = \{\boldsymbol{F}_0\}\sin(\beta t + \vartheta). \tag{6.72}$$

Terms on the left side of the equality in (6.72) are identical in meaning to those described already for (6.66). The forcing function $\boldsymbol{F}$ is the product of a vector of constant amplitudes, $\boldsymbol{F}_0$, and the harmonic multiplier $\sin(\beta t + \vartheta)$, where β is the forcing frequency and ϑ is the phase angle.

Harmonic analysis is used to study the forced vibrational response of a structure under vibrational loads. For example, the method may be exercised to find resonant frequencies and estimate cyclic/fatigue behavior. All loads and displacements are either zero in magnitude or vary sinusoidally (i.e., harmonically) at the same forcing frequency β. Steady-state conditions are assumed, and because superposition principles are invoked, only linear constitutive behaviors are admitted (i.e., linear elastic material response). The intent is estimation of the important features of the response, such as maximum stress or maximum displacement at critical locations in a structure, as a function of loading frequency β.

In FEA of harmonic problems, as is the case for modal analysis of §6.3.2, the geometry and mesh are constructed just as for static problems. Boundary conditions are assigned consistently with (6.72): $\boldsymbol{F}_0$ is specified as nonzero at one or more nodes, and ranges of β and ϑ are prescribed for parametric study. The FE software is used to solve the resulting governing equations. In post-processing, quantities of interest can be resolved versus forcing frequency and phase, providing insight into effects of loading environment on resulting primary and secondary field variables such as deflection and stress, respectively. Again, specific options for solving, post-processing, and associated visualization of results will depend on the features inherently available in the chosen FEM software.

6.4 ANSYS Example: Dynamic Loading

The final numerical example reported in this book demonstrates several concepts from dynamic or fully transient loading with the ANSYS software (v.17.2, 2016). This problem invokes the general theory considered in §6.3.1, meaning the governing equations of 3-D elastodynamics of solid bodies..

6.4.1 Problem Statement

A block or bar of steel rests on a frictionless surface as shown in Fig. 6.2. The elastic properties of this material are treated as isotropic and linear, with modulus $E = 210$ GPa and Poisson's ratio $\nu = 0.3$. The block is of length $L = 1$ m, with a square cross section of dimensions 0.1×0.1 m^2, or area $A = 0.01$ m^2. The mass density is $\rho = 7800$ kg/m^3. A compressive pressure is applied to one side of the block, ramped linearly over time:

$$p = 300\frac{t}{t_0} \text{ MPa}, \quad t_0 = 0.001 \text{ sec.} \tag{6.73}$$

The body is initially at rest and unstressed at $t = 0$. Two conditions are considered for the opposite face. The first includes a rigid fixed backing as shown in the figure. At the boundary between the steel bar and the backing, the displacement is specified as $\boldsymbol{u} = 0$, i.e., all three components are constrained. A second condition to be explored removes the backing entirely, such that the face opposite to that which is applied the pressure is a free surface. In this second case, the entire block or bar is free to accelerate and translate rightwards due to rigid body motion plus stress-induced deformation.

The goals of this example are as follows. First, the time dependent, primary solution for the displacement field is sought for the time domain $0 \le t \le t_0 = 0.001$ sec. Second, time dependent stress contours for the normal stress in the direction of loading are to be obtained for the same time period, providing insight into wave propagation in an elastic solid subjected to ramped loading. In order to establish the proper time incrementation for dynamic analysis involving stress waves, the Courant condition is invoked. Thirdly, wave reflection is explored by consideration of the problem with backing removed, in contrast to static problems considered in

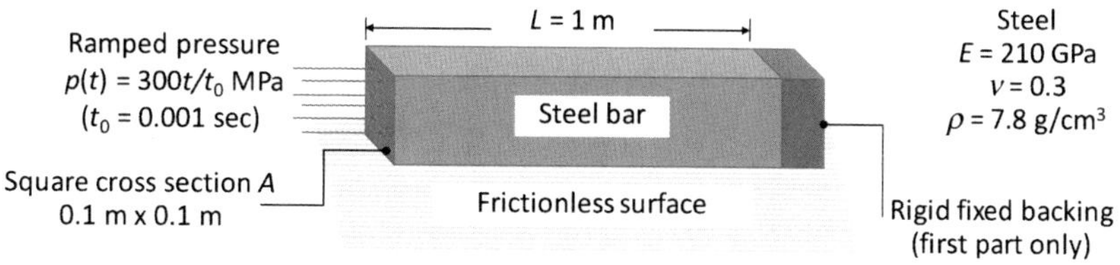

Fig. 6.2 Dynamic example problem: geometry, boundary conditions, and properties

prior chapters in which rigid body motion was prohibited in order to enable a unique numerical solution. In particular, displacement and stress fields for simulations with and without the backing are to be compared.

6.4.2 Pre-processing

Pre-processing consists of the following operations: selection of element type and material properties, geometry creation, mesh generation, and prescription of boundary conditions. The FE analysis in the ANSYS package begins with the choice of an element type for 3-D continuum elasticity, the `tet 10 node 187` element:

```
Main Menu > Preprocessor > Element Type > Add/Edit/Delete
> Add > Tet 10 node 187
```

The constitutive model is isotropic linear elastic with properties invoked by

```
Main Menu > Preprocessor > Material Properties > Material
Models > Structural > Linear > Elastic > Isotropic
```

The elastic modulus E and Poisson's ratio ν are entered as 210e9 Pa and 0.3 as in Fig. 6.2 and Fig. 6.3. The mass density is also necessary for dynamic analysis, here a value of $\rho = 7800$ kg/m^3 in consistent SI units:

```
Main Menu > Preprocessor > Material Properties > Material
Models > Structural > Density
```

Material property data are shown as entered in Fig. 6.3. The simple prismatic geometry is created by the sequence

```
Main Menu > Preprocessor > Modeling > Create > Volumes >
Block > By Dimensions
```

The long edge of the block is parallel to the x-axis. Refer to Fig. 6.4, for example. The analysis type is set to "Transient":

```
Main Menu > Solution > Analysis Type > New Analysis >
Transient
```

The default options of full solution method and no lumped mass approximation are used. Boundary conditions are applied to the faces at $x = 0$ and $x = L$. For the latter, all degrees of freedom are set to zero to simulate a rigid backing:

```
Main Menu > Preprocessor > Define Loads > Apply >
Structural > Displacement > On Areas
```

Pick the face at $x = L = 1$ m and set all degrees of freedom (DOF) to zero. The magnitude of maximum pressure at the end of the analysis is set by imposing a pressure BC on the face at $x = 0$, with the ramping condition to be prescribed later. The maximum pressure is set via the sequence

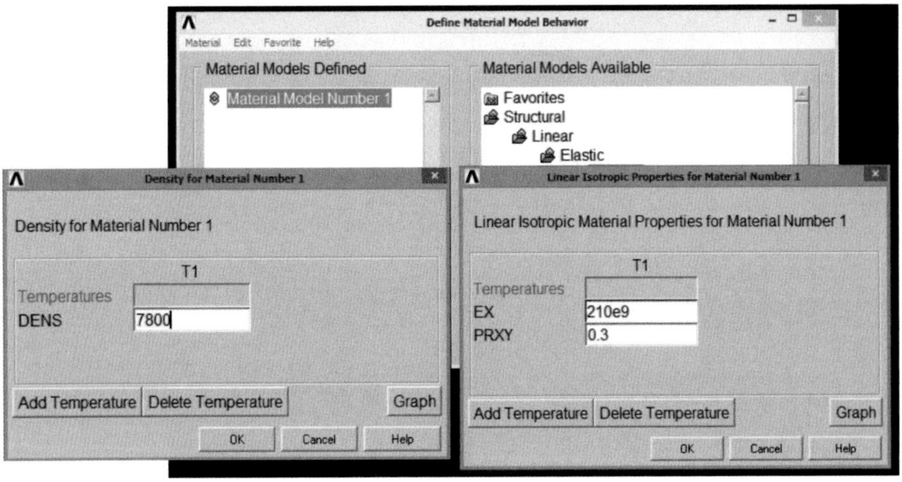

Fig. 6.3 Dynamic example problem: material properties

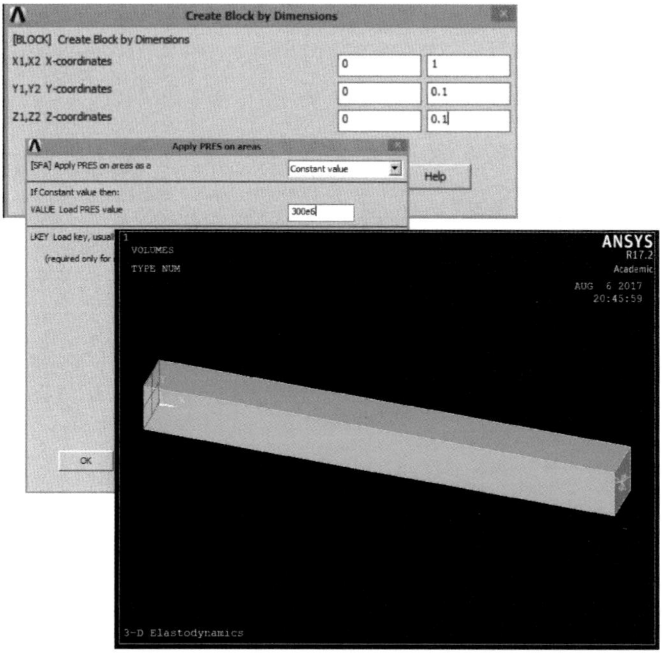

Fig. 6.4 Dynamic example problem: geometry and boundary conditions

```
Main Menu > Preprocessor > Define Loads > Apply >
Structural > Pressure > On Areas
```

Pick the face at $x = 0$ and enter a pressure value of 300 MPa in accordance with Fig. 6.2. Next, the FE mesh is generated. The mesh size is set by

```
Main Menu > Preprocessor > Meshing > Size Cntrls > Manual
Size > Global > Size
```

A value of 0.05 is entered for the element edge length. The mesh in Fig. 6.5 is then produced by the meshing command:

```
Main Menu > Preprocessor > Mesh > Volumes > Free > Pick
```

Choose the only existing volume and then click “OK”.

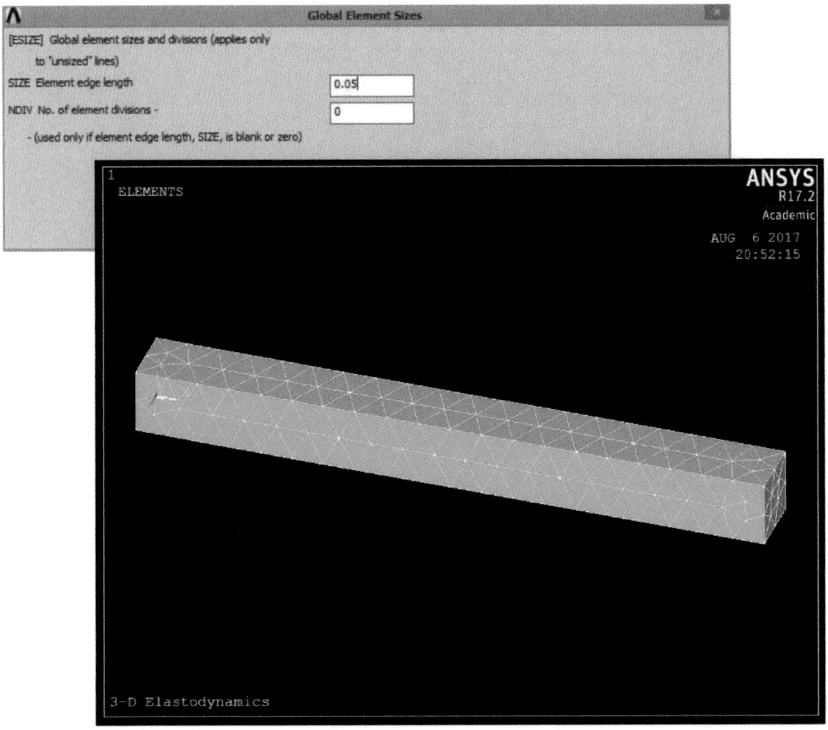

Fig. 6.5 Dynamic example problem: tetrahedral mesh

6.4.3 Solution

The transient solution is obtained after selection of some solver parameters. With an element length of 0.05 m, the Courant condition in (6.58) suggests that the maximum time step size should not exceed

$$(\Delta t)_{\max} \leq \frac{L_{\min}}{C_s} \approx \frac{0.05 \text{ m}}{5190 \text{ m/s}^2} \approx 1 \times 10^{-5} \text{sec}, \tag{6.74}$$

where the stress wave speed C_s is approximated as $\sqrt{E/\rho}$. This calculation is used to guide choices of minimum and maximum time steps in the solution controls:

```
Main Menu > Solution > Analysis Type > Sol'n Controls
```

In the "Basic" tab, the time step sizes and the schedule for writing of solution output are specified, while in the "Transient" tab, the ramped loading condition is enforced, consistent with the imposed pressure increasing linearly in time to its maximum at $t = t_0$. The requisite fields are shown in Fig. 6.6. Notice that the end time of

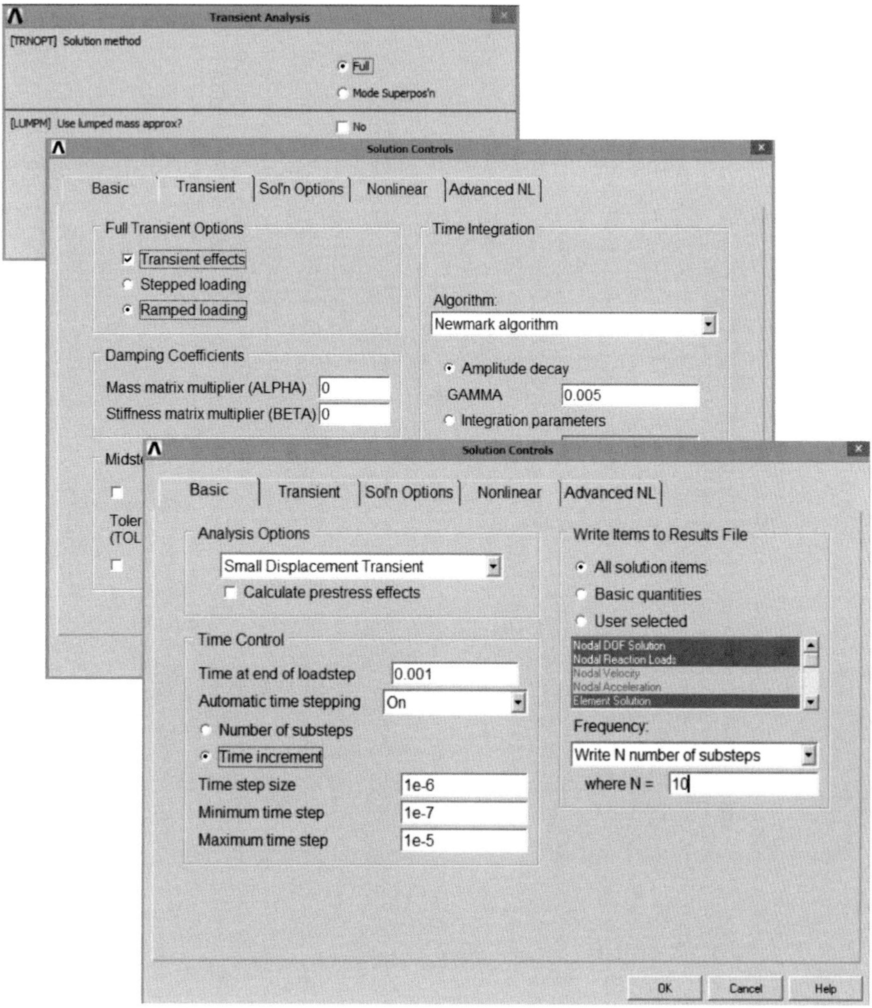

Fig. 6.6 Dynamic example problem: solution controls

the simulation is $t = t_0 = 0.001$ sec. The transient solution for the time domain $t \in [0, 0.001]$ sec is then obtained by the usual sequence

```
Main Menu > Solution > Solve > Current LS > OK
```

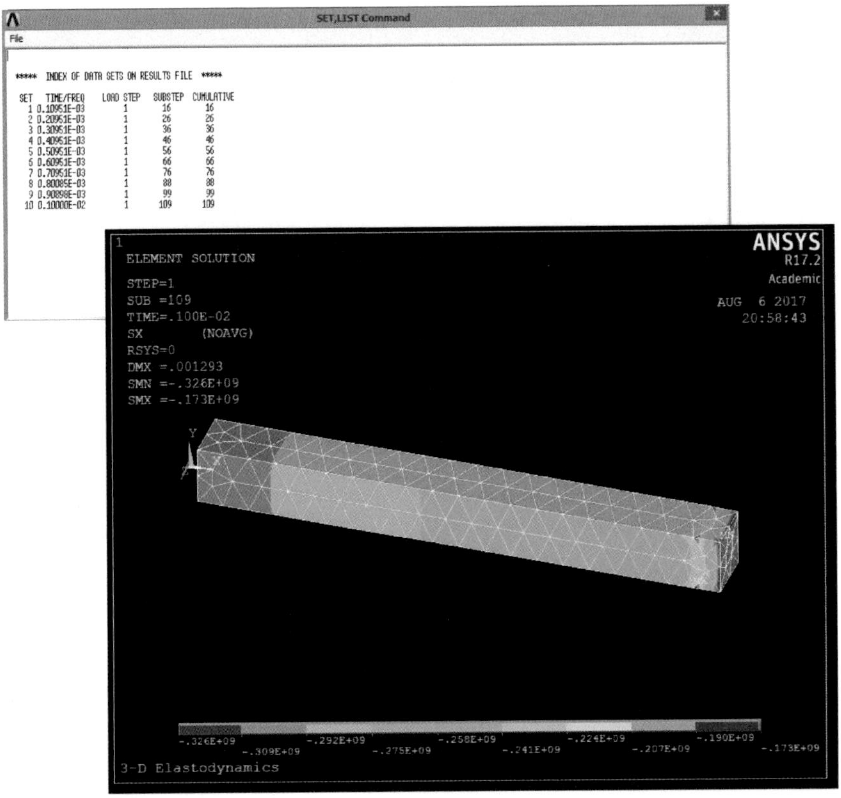

Fig. 6.7 Dynamic example problem: axial stress contour in elements, $t = 0.001$ sec, rigid backing

6.4.4 Post-processing

The deformed shape of the body, the maximum displacement over all nodes in the body, and the distribution of axial stress σ_{xx} in the body are sought for this example problem. First, the deformed structure at the very end of the time period can be viewed in superposition with the original body via

```
Main Menu > General Postproc > Plot Results > Deformed
Shape
```

Choose "Deformed and Undeformed Edge". The displacements are magnified by the ANSYS GUI to facilitate visualization. A contour plot of axial stress in the elements is created by the sequence below:

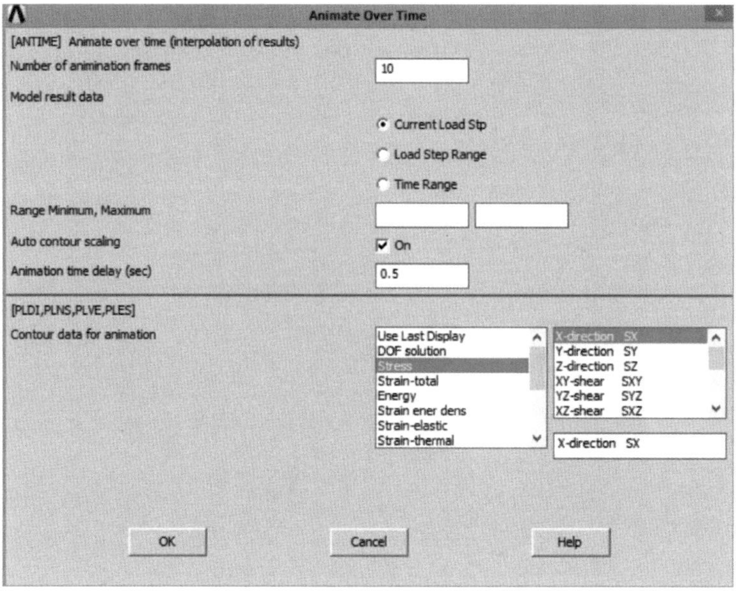

Fig. 6.8 Dynamic example problem: animation settings

```
Main Menu > General Postproc > Plot Results > Element
Solution > X-Component of stress
```

The result is Fig. 6.7, which also shows the time step and some numerical integration information obtained via

```
Main Menu > General Postproc > Results Summary
```

Videos and snapshots in time for other instances in the analysis ($t \leq 0.001$ sec) can be obtained as follows. As shown in Fig. 6.8, animation is enacted by

```
Utility Menu > PlotCtrls > Animate > Over Time
```

The nodal solutions obtained by taking static images of the results for the axial stress field at $t \approx 0.0005$ sec and at the end time of 0.001 sec are shown in Fig. 6.9. Maximum compressive axial stresses are approximately equal to the maximum applied pressure, with a magnitude on the order of 300 MPa.

The second sub-problem, in which the rigid fixed backing in Fig. 6.2 is removed, is now addressed. The essential BC on displacement is removed at the face corresponding to $x = L = 1$ m:

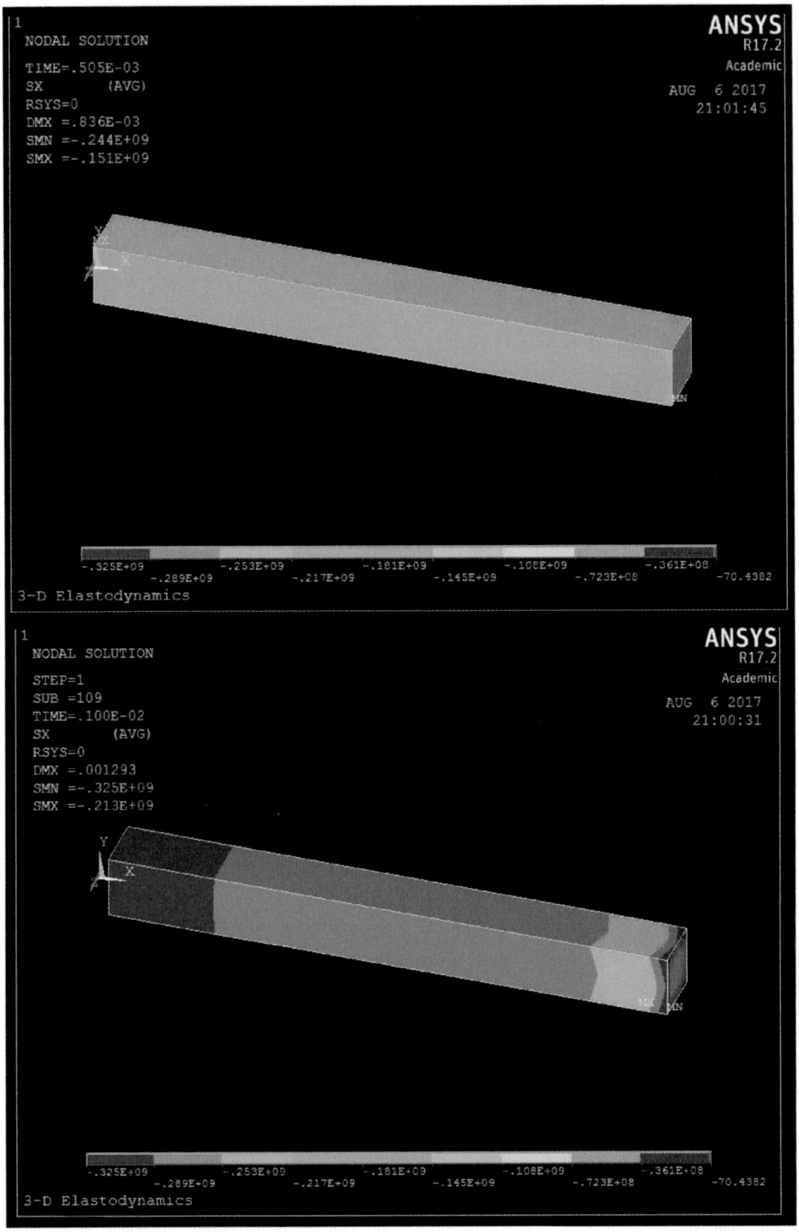

Fig. 6.9 Dynamic example problem: nodal solution for axial stress contours, $t \approx 0.0005$ sec (top) and $t = 0.001$ sec (bottom), rigid backing

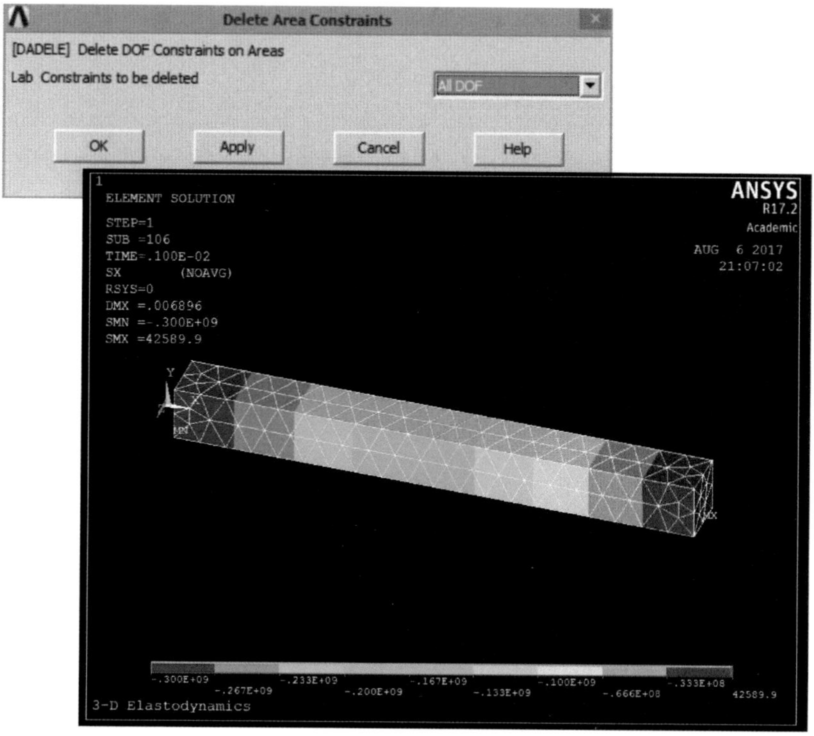

Fig. 6.10 Dynamic example problem: axial stress contour in elements, $t = 0.001$ sec, no backing

```
Main Menu > Preprocessor > Loads > Define Loads > Delete >
Structural > Displacement > On Areas > Pick
```

See Fig. 6.10. Choose the correct area and click "OK". The remaining parameters in the analysis are unchanged, and the solution proceeds via

```
Main Menu > Solution > Solve > Current LS > OK
```

Post-processing steps are followed as for the first sub-problem. The element solution shown in Fig. 6.10 reports σ_{xx} at the end time of the analysis. The nodal solutions shown in Fig. 6.11 report σ_{xx} at the midpoint ($t \approx 0.0005$ sec) and the end time $t = 0.001$ sec. Differences resulting from the BC, or lack thereof, at the right side of the block are evident. When the rigid BC is enacted, the incident compressive wave reflects off of the constrained back surface as a compression wave, leading to more strongly negative pressure. When the rigid BC is removed, the incident compressive wave reflects off of the free back surface as a tensile wave, reducing the compressive pressure. In the former first case, the maximum displacement is

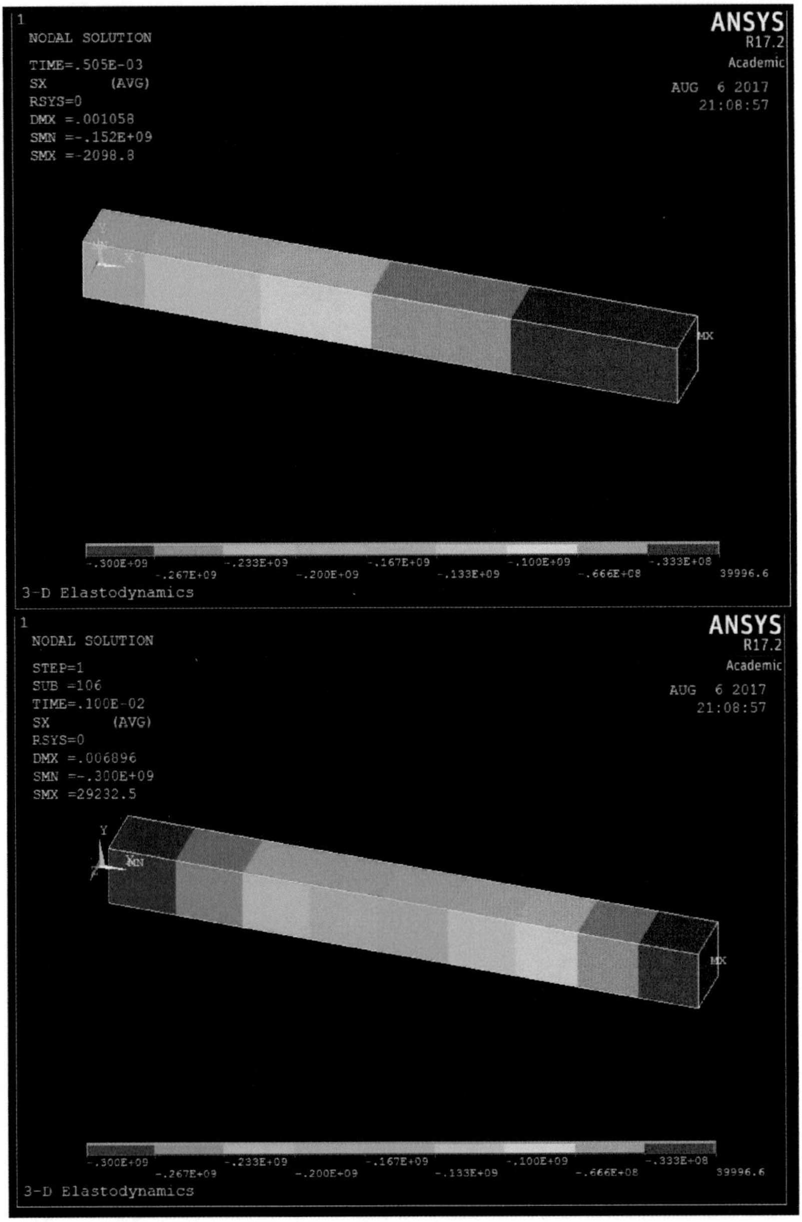

Fig. 6.11 Dynamic example problem: nodal solution for axial stress contours, $t \approx 0.0005$ sec (top) and $t = 0.001$ sec (bottom), no backing

impeded by the rigid BC, which inhibits the block from moving from left to right. In the latter second case, the block is able to translate rigidly in the x-direction.

The displacement of the center of mass of the block can be computed analytically when the body is treated as perfectly rigid. The mass M of the block is the density times the volume:

$$M = \rho LA = 7800 \cdot 1 \cdot 0.01 \text{ kg} = 78 \text{ kg}. \tag{6.75}$$

From Newton's second law, the acceleration $a(t)$ is the total axial force $p(t) \cdot A$ divided by the mass:

$$a(t) = \frac{p(t)A}{M} = \frac{3 \times 10^6}{78} \frac{t}{t_0} = 38500 \frac{t}{t_0} \text{ m/s}^2. \tag{6.76}$$

Integrating the acceleration twice with respect to time and noting that the initial velocity and initial displacement are both zero, the displacement function $u(t)$ becomes

$$u(t) = \frac{1}{6t_0} \cdot 38500 t^3 = 6.41 \times 10^6 t^3 \text{ m}. \tag{6.77}$$

At $t = t_0 = 0.001$ sec, the rigid block is predicted to have translated a distance of 6.41×10^{-3} m. This value is substantial compared to the (maximum) displacement induced by elastic deformation and wave mechanics, much larger than the value of 1.3×10^{-3} m obtained from ANSYS when the backing is included (Fig. 6.9, bottom), and it is of the same order as the value of 6.9×10^{-3} m when it is removed (Fig. 6.11, bottom). Thus, we conclude that much of the motion incurred in the second sub-problem, when the backing is absent, is due to rigid translation.

6.5 Problems

6.5.1. Consider a thick column made of an isotropic, linear elastic solid material as shown in Fig. 6.12. The length of the column is 10 mm and the radius of the column is 2 mm. The column is clamped at the center of one end to prevent rigid body motion, and it is free to expand laterally but not axially. A distributed pressure load of magnitude $p = 1$ GPa is applied normal to the opposite face of the column. [Note:

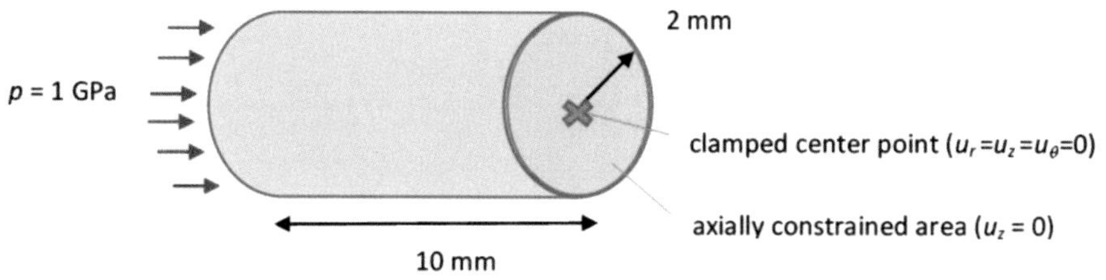

Fig. 6.12 Problem 6.5.1 (see also Problem 5.9.1)

this is the same geometry studied in (quasi-static) Problem 5.9.1 and Fig. 5.9.] Assume that p is applied instantly at time $t = 0$, as a step function (i.e., Heaviside function with respect to time).
a. First take the column to be made of the solid metal lead with mass density $\rho = 11.3$ g/cm^3, elastic modulus $E = 16$ GPa, and Poisson's ratio $\nu = 0.45$. Using finite element software, analyze the problem for fully dynamic loading conditions invoking either a fully 3-D model or an axisymmetric model. Describe the steps used to set or create the geometry, initial and boundary conditions, and time incrementation protocols.
b. Obtain figures showing the deformed shape of the cylinder and axial stress contour(s) at $t = 5\mu$s.
c. Repeat the steps of part **b.** using the following material properties for diamond: $\rho = 3.5$ g/cm^3, $E = 1100$ GPa, and $\nu = 0.07$. Does the compression wave appear to travel faster in lead or in diamond, and why?

6.5.2. Consider a homogeneous plate in the shape of an equilateral triangle as shown in Fig. 6.13, with all three edges clamped. A clamped edge is constrained such that all degrees of freedom—all displacements and rotations—are zero on nodes located on the edges. Material properties of the plate are $E = 1$ GPa, $\nu = 0.3$, and $\rho = 1$ g/cm^3. The plate's thickness is 0.02 m, and its edge length is $L = \sqrt{3}H = 1$ m.
a. Perform modal analysis of the plate using FE software. Create a mesh of quadrilateral shell elements with approximately 1000 nodes.
i. Determine and list the first (i.e., lowest) 20 natural frequencies. Explain the steps and parameters that you used.
ii. Obtain plots of the deformed mesh showing the shapes of the 2nd, 10th, and 19th modes.
b. Perform harmonic analysis of the plate using the same mesh as in part **a.**, presumably with the same software package. Apply a force at a point p on the triangle face located at the centroid, or to within a very small distance from the centroid as mesh tolerances permit; see Fig. 6.13. Apply a force of magnitude $F_0 = 10$ N in the direction normal to the plane of the plate. Consider a range of forcing frequencies β with null phase angle, i.e., $\vartheta = 0$ in equation (6.72) of the main text. The three

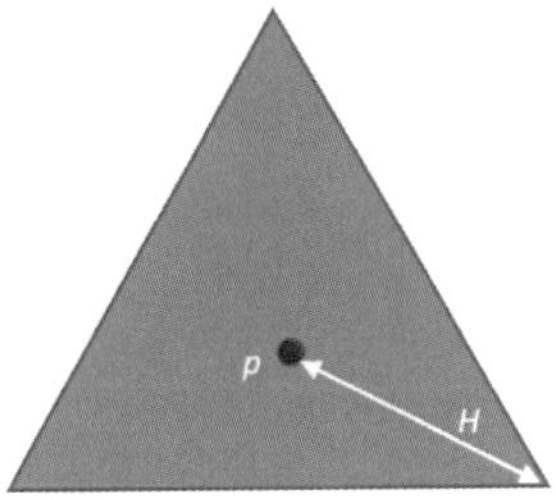

Fig. 6.13 Problem 6.5.2

edges of the plate remain clamped throughout the time duration of the problem, as in part **a.**
i. Choose an appropriate set of numerical parameters for the analysis and provide an explanation justifying your choices.
ii. Report the first 5 resonant peaks, specifically the frequency β and maximum out-of-plane deflection for each. Plot the frequency versus maximum out-of-plane displacement curve for point p.

References

1. L.E. Malvern, *Introduction to the Mechanics of a Continuous Medium* (Prentice-Hall, Englewood Cliffs, NJ, 1969)
2. J.D. Clayton, *Nonlinear Mechanics of Crystals* (Springer, Dordrecht, 2011)
3. J.D. Clayton, Dynamic plasticity and fracture in high density polycrystals: constitutive modeling and numerical simulation, *Journal of the Mechanics and Physics of Solids* **53**, 261–301 (2005)

Glossary

Important terminology is summarized in the alphabetized glossary entries below:

accuracy proximity of an approximate solution to the exact or true solution of a given problem

assembly combination of local vectors and/or matrices for each finite element into a global vector and/or matrix

axisymmetric problem geometry that can be described using cylindrical coordinates wherein field variables do not depend on angular position about an axis of rotational symmetry

bar element one-dimensional element for solid mechanics supporting only axial loads

beam element one-dimensional element (with 2-D deformation) for solid mechanics supporting transverse loads and bending moments

boundary conditions prescribed constraints on values of primary or secondary variables at regions or endpoints of a problem domain enabling a non-trivial solution

compatibility a mathematical relationship, often stated in the form of differential equations and often of kinematic origin, between primary and certain secondary variables in an analysis

complete polynomial a polynomial function containing all terms up to its highest order

condensed system stiffness matrix, degrees-of-freedom vector, and/or load vector for a global system of governing equations modified to account for essential boundary conditions

connectivity shared nodes among elements in a finite element mesh

conservation law a fundamental governing equation of physics whose solution is sought in an analysis

constitutive model a mathematical representation of material behavior, including any associated parameters or constants specific to a material or class of materials

continuum field theory any branch of physics wherein primary and secondary variables are mathematically represented by continuous fields over the domain rather than discrete values at a finite set of points within that domain

direct solver method of solution involving exact inversion of stiffness matrix

Dirichlet boundary condition see essential boundary condition

elasticity a continuum field theory of solid mechanics wherein the response of the material obeys a certain mechanically, if not fully thermodynamically, reversible relationship between stress and strain

element subregion of problem domain obtained by discretization

essential boundary condition imposed value or constraint on a primary field variable; also called Dirichlet boundary condition

Euler-Bernoulli beam theory mechanical theory for beams that assumes planar cross sections remain planar and normal to the neutral axis, omits shear strains, and invokes linear elastic constitutive behavior

finite element analysis (FEA) see finite element method

finite element method (FEM) technique for solving differential equation(s) of a boundary value problem whereby domain is discretized into element(s) and resulting algebraic equation(s) yield values of primary variable(s) at nodes

finite element software computer code facilitating analysis of problems via the finite element method, often consisting of modules for pre-processing, solving, and post-processing

Fourier conduction constitutive model for heat conduction wherein the heat flux vector is proportional, via the thermal conductivity matrix/tensor, to the spatial temperature gradient

frame element two-dimensional finite element for solid mechanics that combines loads and degrees of freedom from truss/bar and beam elements

free mesh discretization, typically performed by FEM software, without significant pre-imposed constraints

harmonic analysis study of a time dependent physics problem wherein boundary conditions, e.g., applied loads and displacements for a solid mechanics problem, are sinusoidal with respect to time

Hermite shape functions interpolation functions enabling description of a variable and its derivative; used for beam elements to interpolate transverse deflection and its spatial derivative, the slope

h-refinement improvement in mesh resolution via increasing the local or global mesh density

invertible matrix matrix whose inverse exists and is necessarily non-singular

interpolation function see shape function

iterative solver method of solution involving approximate inversion of stiffness matrix via sequential computations

mapped mesh discretization with parameters or constraints imposed by the user

mesh a discretization or grid of elements covering a domain in FEA

modal analysis study of a time dependent physics problem involving solution of an eigenvalue problem wherein boundary conditions, e.g., applied loads and displacements for a solid mechanics problem, are of zero magnitude

natural boundary condition imposed value or constraint on a secondary field variable; also called Neumann boundary condition

Neumann boundary condition see natural boundary condition

node point in discretized problem domain at which value of a primary field variable is sought

orthogonal matrix matrix whose transpose is equal to its inverse; a rotation matrix for coordinate transformation is a common example

plane strain solid mechanics boundary conditions corresponding to null out-of-plane strain components in a two-dimensional analysis

plane stress solid mechanics boundary conditions corresponding to null out-of-plane stress components in a two-dimensional analysis

Poisson's ratio dimensionless elastic stiffness coefficient often relating orthogonal stress/strain components

p-refinement improvement in mesh resolution via increasing the order of interpolation function(s)

primary variable field variable whose value is obtained directly in an analysis by solution of some version of the governing equation(s)

secondary variable field variable whose value is obtained in an analysis from mathematical manipulation of primary variable(s)

semi-discrete equation a partial differential equation with two or more independent variables where discretization is applied with respect to some, but not all, of the independent variables

shape function mathematical function used to interpolate FEM solution at nodal points to other locations within an element

singular matrix matrix whose determinant has value zero

skew matrix matrix that is equal to the negative of its transpose; also called anti-symmetric

sparse matrix matrix with a large fraction of null or zero entries

stability boundedness of a (numerical) solution

statically indeterminant system of equations without a solution; in FEA, a condition often associated with lack of sufficient constraint, rigid body modes, and/or a singular global stiffness matrix

stress a directional force per unit area; a secondary solution variable in solid mechanics problems

strong form complete statement of a boundary value problem in terms of original differential equation(s) and supplementary information

symmetric matrix matrix that is equal to its transpose

transient problem a physical problem whose solution varies with time; also may be labeled a dynamic problem

tridiagonal matrix matrix that has zeroes for most entries except those entries on its diagonal and those entries immediately neighboring its diagonal

truss element one-dimensional element for solid mechanics

validation determination if a solution to a mathematical representation of a problem is physically correct

verification determination if a numerical solution to a problem is mathematically correct or mathematically accurate

weak form complete statement of a boundary value problem in terms of weakened (integro-)differential equation(s) derived from the strong form and possible supplementary information

Young's modulus elastic stiffness coefficient; ratio of stress to strain for uniaxial load problem in linear elasticity

List of Symbols

Notation such as that for mathematical symbols used throughout the text is listed below. This list is not all-inclusive; other notation may be used at times and is defined where first encountered in the main chapters.

Item	Definition (equation of first appearance)
$\boldsymbol{K}$	stiffness matrix (1.1)
$\boldsymbol{u}$	primary field vector (e.g., displacement) (1.1)
$\boldsymbol{F}$	load vector (1.1)
k	spring constant (2.12) or isotropic thermal conductivity (4.7)
E	elastic modulus or Young's modulus (2.19)
A	cross-sectional area (2.19)
L	length of a finite element (2.19)
$u_j^{(e)}$	displacement vector component for element e and local node j (2.25)
$Q_j^{(e)}$	force vector component for element e with local node j (2.25)
$k_{ij}^{(e)}$	stiffness matrix component for element e with indices i, j (2.25)
$\boldsymbol{T}^{(e)}$	coordinate transformation matrix (2.50)
σ	axial stress (2.73)
ε	axial strain (2.73)
ε_e	elastic strain (2.86)
ε_t	thermal strain (2.86)
α	coefficient of thermal expansion (2.86) or rotation angle (3.94)
T	temperature (2.86)
θ	rotation angle for coordinate transformation (2.50) or beam rotation (3.2)
w	weight function or test function for determining the weak form (2.103)
N_i	shape function or interpolation function (2.107)
$\boldsymbol{B}$	matrix of shape function derivatives (2.108)
δ_{ij}	Kronecker delta (2.109)
υ	beam deflection (3.1)
κ	beam curvature (3.4)
M	bending moment (3.7)

I	moment of inertia (3.7)
V	shear force (3.9)
$\boldsymbol{q}$	heat flux vector (4.1)
f	scalar heat source (4.1)
$\boldsymbol{k}$	thermal conductivity matrix (4.3)
Ω	domain of integration (4.9)
$\partial\Omega$	boundary of domain of integration (4.9)
q_n	scalar heat flux (4.17)
A_e	area of a 2-D or axisymmetric finite element (4.20)
ε_{ij}	strain tensor (4.50)
σ_{ij}	stress tensor (4.54)
C_{ijkl}	elasticity tensor or elastic stiffness tensor (4.58)
W	strain energy density (4.60)
$\boldsymbol{t}$	traction vector (4.62)
ν	Poisson's ratio (4.65)
σ_v	von Mises or effective stress (4.96)
λ	Lamé modulus (5.62)
G	shear modulus (5.62)
e	elastic dilatation (5.63)
ω	elastic rotation (5.63)
ρ	mass density (6.2)
c	specific heat capacity (6.2)
$M_{ij}^{(e)}$	mass matrix component for element with indices i, j (6.24)
$\boldsymbol{a}$	acceleration vector (6.28)
C_s	signal velocity or wave speed (6.58)

Operator	Description
$\cdot$	dot product
$\times$	cross product
$:$	double dot product
$\otimes$	tensor (outer) product
$\ln(\cdot)$	natural logarithm
$\det(\cdot)$	determinant of square matrix or second-order tensor
$\mathrm{tr}(\cdot)$	trace of second-order tensor
$(\cdot)^{-1}$	inverse of a function or matrix
$(\cdot)^{\mathrm{T}}$	transpose of matrix or second-order tensor
$\nabla(\cdot)$	spatial gradient
$\nabla^2(\cdot)$	spatial Laplacian
$\partial(\cdot)/\partial t$	(partial) time derivative
$\int(\cdot)$	generic volume, surface, or line integral
$\oint(\cdot)$	integral over a closed surface or a closed curve
$\bigwedge$	assembly acting over elements

Solutions

Solutions to selected problems are given below. Solutions are not provided for all parts of all problems contained in the main text, and in some cases only partial solutions or hints are given to aid the student/reader.

Problems of Chapter 2

2.6.1

a. Hint: demonstrate $N_i(x_j) = \delta_{ij}$ for $i, j = 1, 2, 3$.

b. $[\boldsymbol{B}] = \dfrac{\mathrm{d}[\boldsymbol{N}]}{\mathrm{d}x} = \dfrac{1}{2L^2}[2x - L \quad -4x \quad 2x + L]$.

c. $[\boldsymbol{k}^{(e)}] = \dfrac{EA}{6L}\begin{bmatrix} 7 & -8 & 1 \\ -8 & 16 & -8 \\ 1 & -8 & 7 \end{bmatrix}$.

d. Hint: compare with equation (2.31) of Chapter 2.

2.6.2

a. Hint: global stiffness matrix should be 4×4; see equation (2.37) of Chapter 2.

b. Hint: condensed stiffness matrix should be 3×3 since $u_1 = 0$; see equation (2.43).

c. $\{\boldsymbol{u}\} = [u_1 \quad u_2 \quad u_3 \quad u_4]^{\mathrm{T}} = 10^{-7} \cdot [0 \quad 8 \quad 32 \quad 40]^{\mathrm{T}}$.

d. Forces in elements 1, 2, 3, 4, 5 = 20, 10, 30, 10, 20 kN.

e. Stresses in MPa = 0.5, 1, 3, 1, 0.5; strains $\times 10^6$ = 4, 4, 12, 4, 4.

Problems of Chapter 3

3.9.1

a. Two elements. To build a global stiffness matrix, see Fig. 3.6 in Chapter 3, giving

$$[\boldsymbol{K}] = \frac{2EI}{L_e^3}\begin{bmatrix} 6 & -3L_e & -6 & -3L_e & 0 & 0 \\ -3L_e & 2L_e^2 & 3L_e & L_e^2 & 0 & 0 \\ -6 & 3L_e & 12 & 0 & -6 & -3L_e \\ -3L_e & L_e^2 & 0 & 4L_e^2 & 3L_e & L_e^2 \\ 0 & 0 & -6 & 3L_e & 6 & 3L_e \\ 0 & 0 & -3L_e & L_e^2 & 3L_e & 2L_e^2 \end{bmatrix},$$

where $L_e = 5$m and $EI = 10^9$N·m^2.
b. Assembled force vector is

$$\{\boldsymbol{F}\} = -\frac{f_0}{6}\begin{Bmatrix} 10 \\ -10 \\ 10 \\ 10 \\ 0 \\ 0 \end{Bmatrix} + \begin{Bmatrix} Q_1^1 \\ Q_2^1 \\ Q_3^1 + Q_1^2 \\ M \\ -F - ku_5 \\ 0 \end{Bmatrix},$$

where $f_0 = 3 \cdot 10^6$N/m, $M = 7 \cdot 10^5$N·m, $F = 4 \cdot 10^6$N, and $k = 7 \cdot 10^5$N/m. Distributed load f contributes only to forces for element 1.
c. Essential boundary conditions are $u_1 = u_2 = u_3 = 0$. Condensed system is

$$\frac{2EI}{L_e^3}\begin{bmatrix} 4L_e^2 & 3L_e & L_e^2 \\ 3L_e & [6 + kL_e^2/(2EI)] & 3L_e \\ L_e^2 & 3L_e & 2L_e^2 \end{bmatrix}\begin{Bmatrix} u_4 \\ u_5 \\ u_6 \end{Bmatrix} = \begin{Bmatrix} -5f_0/3 + M \\ -F \\ 0 \end{Bmatrix}.$$

d. Primary solution is

$$\begin{Bmatrix} u_4 \\ u_5 \\ u_6 \end{Bmatrix} = \begin{Bmatrix} 0.0185 \text{ rad} \\ -0.2518 \text{ m} \\ 0.0663 \text{ rad} \end{Bmatrix}.$$

e. Bending moments M_i and shearing forces V_i at global nodes i obtained using B-matrix and derivative of B-matrix:

$$\begin{Bmatrix} M_1 \\ M_2^{(\text{el. }1)} \\ M_2^{(\text{el. }2)} \\ M_3 \end{Bmatrix} = \begin{Bmatrix} 7.4 \\ -14.8 \\ -19.1 \\ 0 \end{Bmatrix} \text{MN} \cdot \text{m}, \qquad \begin{Bmatrix} V_1 \\ V_2^{(\text{el. }1)} \\ V_2^{(\text{el. }2)} \\ V_3 \end{Bmatrix} = \begin{Bmatrix} -4.4 \\ -4.4 \\ 3.8 \\ 3.8 \end{Bmatrix} \text{MN}.$$

f. Using the B-matrix gives $M(x = 7.5\text{m}) = -9.6$ MN·m.

3.9.2

b. $\upsilon(x) = \frac{f_0 x^2}{24EI}\left(6L^2 - 4Lx + x^2\right)$ where $L = 1$m, $E = 10^{11}$N/m^2, $f_0 = 10^4$N/m, $I = \frac{bh^3}{12}$, $b = h = 0.1$m. Deflection at $x = L$ is 0.0015m; rotation $\theta = -\mathrm{d}\upsilon/\mathrm{d}x$ at $x = L$ is 0.002°.

Problems of Chapter 4

4.11.1

a. See equations (4.20)–(4.25) in Chapter 4.
b. Matrix of coefficients with area pre-factor and local properties are

$$\frac{1}{2A}\begin{bmatrix}\alpha_1 & \alpha_2 & \alpha_3\\ \beta_1 & \beta_2 & \beta_3\\ \gamma_1 & \gamma_2 & \gamma_3\end{bmatrix} = \frac{1}{2\cdot 3}\begin{bmatrix}6 & 0 & 0\\ -2 & 2 & 0\\ -1 & -2 & 3\end{bmatrix}; \qquad N_1(0,0) = N_2(3,0) = N_3(2,2) = 1.$$

Each shape function has a value of zero at the other two nodes.

4.11.2

b. A temperature field that appears visually sensitive to mesh construction is a possible, but not necessary, condition that the mesh from part **a.** is insufficiently fine.
c. Solutions do not depend on k for steady conduction in the absence of point sources, as is clear from equation (4.8) of §4.1.

Problems of Chapter 5

5.9.1

a. and **b.** Validation of computer results includes near-uniform axial strain $\varepsilon_{zz} \approx -\sigma_{zz}/E = -0.0625$ and maximum (in magnitude) axial displacement $u_z = \upsilon = \varepsilon_{zz} \cdot L \approx -0.625$ mm.

Problems of Chapter 6

6.5.1

c. Compression waves travel much faster in diamond than in lead since the longitudinal wave speed in a linear elastic solid is $C_s \approx \sqrt{E/\rho}$.

6.5.2

a. First/lowest natural frequency is ≈ 667 Hz; twentieth is ≈ 5840 Hz.
b. First resonant frequency should be very close to first natural frequency.

Index

Made in the USA
Middletown, DE
28 January 2020

83805261R10119